From the Ground Up

Also by Charles Wing

From the Walls In

Also by John N. Cole

In Maine
Striper

From the Ground Up

JOHN N. COLE
and
CHARLES WING

Illustrations by Tom Paiement

An Atlantic Monthly Press Book
Little, Brown and Company   Boston Toronto

FIRST EDITION

E

Atlantic-Little, Brown Books
are published by
Little, Brown and Company
in association with
The Atlantic Monthly Press

Designed by Susan Windheim
*Published simultaneously in Canada
by Little, Brown & Company (Canada) Limited*

PRINTED IN THE UNITED STATES OF AMERICA

MV

Library of Congress Cataloging in Publication Data

Cole, John N   1923-
    From the ground up.

    "An Atlantic Monthly Press book."
    Includes index.
    1. House construction—Amateurs' manuals. I. Wing, Charles,
1939-   joint author. II. Title.
TH4815.C65   690'.8   76-14825
ISBN 0-316-15112-2

Table 1 and Illustration 23 are reprinted with permission from
Architectural Graphic Standards, edited by C. G. Ramsey and
H. R. Sleeper, 6th ed., John Wiley and Sons, New York, 1970.

Illustrations 18, 19, 20, 70, 71, 112, and 114 are reprinted
with permission from *Handbook of Air Conditioning, Heating
and Ventilating,* 2nd edition, Strock and Koral, published by
Industrial Press Inc., New York.

Illustrations 43 and 110 are from *Time Saver Standards,*
edited by John Hancock Callendar. Copyright © 1974, 1966 by
McGraw-Hill, Inc. Used with permission of McGraw-Hill Book
Company.

Illustration 122 is used with permission of the supplier of the
graph, Jøtul, and Kristia, translator and rewriter.

For the scores of shelter builders who helped with ours,
and for those who will help you with yours.
Without them, there would be no shelters.

Contents

Introduction

There is some of the same fitness in a man's building his own house that there is in a bird's building its own nest. Who knows but if men constructed their dwellings with their own hands and provided food for themselves and families simply and honestly enough, the poetic faculty would be universally developed, as birds universally sing when they are so engaged? But alas! we do like cowbirds and cuckoos, which lay their eggs in nests which other birds have built, and cheer no traveller with their chattering and unmusical notes. Shall we forever resign the pleasure of construction to the carpenter? What does architecture amount to in the experience of the mass of men? I never in all my walks came across a man engaged in so simple and natural an occupation as building his house. We belong to the community. It is not the tailor alone who is the ninth part of a man: it is as much the preacher, and the merchant, and the farmer. Where is this division of labor to end? And what object does it finally serve? No doubt another *may* also think for me; but it is not therefore desirable that he should do so to the exclusion of my thinking for myself.

—THOREAU, *Walden*

This book began with one house, then two, then more.

One house was built by one of the authors, John Cole, another by the other author, Charlie Wing. Both houses were built from the ground up, both were built in Maine, and, in due course, both became the topics of shared conversations between John and Charlie, their families and their friends.

As a professor of physics at Bowdoin College (Brunswick, Maine), Charlie Wing also began teaching a course in shelter design and building. It was a natural progression from his two careers — the first as a professor, the second as a shelter builder.

As a newspaper editor (*Maine Times*), John Cole began writing and editorializing about his avocation of 1971 and '72 — the building of a shelter by and for the considerable Cole family.

Charlie's course was one of the most popular at Bowdoin; Cole's articles about his shelter progress drew more mail than most *Maine Times* articles.

Too many students for Charlie; too much mail for John.

Out of the resulting confusion came the beginnings of a plan. A need existed, John and Charlie agreed, for full-time schools where students of all ages could learn how to plan, design and build their own low-cost post-industrial homes. Also needed were places where shelter information could be collected, expertly evaluated, and shared.

In 1973 Charlie co-founded The Shelter Institute. Charlie continued teaching while his partners ran the business. In mid-1976 Charlie severed his association with lawyer Patrice Hennin and started Cornerstones: The Wing School of Shelter Technology in Brunswick, Maine.

Once Charlie realized he had written down most of his lectures, the notion of this book became inevitable.*

At that point, sometime in late 1973, John and Charlie were helped toward the solution by Pat and Patsy Hennin, who came along and, with Charlie, created and opened a shelter building school and resource center. They named it The Shelter Institute. The original school continues under the Hennins. Charlie now teaches shelter building at Cornerstones owner building school in Brunswick, ME. He also teaches there retrofitting of old houses and is working on a book on the same subject.

Once the institute was organized and began to collect a sizable amount of detailed information, shelter wisdom, and experience, and once Charlie realized he had written down most of his classroom lectures, which had been refined over a period of three years, the notion of this book became inevitable.

Charlie had to be one of the authors — he had developed the body of knowledge. John Cole wanted to be one of the authors because he believes so emphatically in the need for what he calls the post-industrial ethic. So they wrote this

---

*In accordance with the terms of Charlie's separation from Shelter Institute, we print the following: "written with the cooperation of the Shelter Institute, Inc., Bath, Maine."

book together, in alternate chapters (or close to it) like a layer cake: a bit of icing (Cole) here; a good deal of nourishment (Wing) there. Or, to put it another way, romanticist Cole wrote the metaphysics; scientist Wing wrote the physics.

Although the Coles and the Wings have never discussed a detailed definition of the term "post-industrial," each of them would agree with whatever definition the others might apply. They would because the term is more of a feeling than a dogma, more of a philosophical poem than a tautological dictionary.

That philosophy involves the notion that this nation (and others) is making a transition from the industrial-age ethic of mass production, mass consumption, and mass waste to a "post-industrial" ethic of community production, careful consumption, and almost no waste.

A parallel transition also in progress involves the fundamental realization that rather than attempting to utilize science and technology to "conquer" nature, man must instead use those assets to live in greater harmony with nature.

The major catalyst precipitating the transition from industrialism to post-industrialism is the growing global recognition of the finiteness of nonrenewable resources — oil, coal, copper, steel, aluminum, and the like. Once the general populace understands (and they do) that this planet can run out of nonrenewables, the entire concept of continued and perpetual growth must be corrected. (If we run out of the resources that fuel growth, how can growth continue? If the planet can produce just so much food, then there will have to be just so many people.)

From these blunt truths springs the reality of equilibrium systems as opposed to growth systems, and that's what post-industrialism is all about. If you think about it a bit, you understand why an equilibrium system pivots on community production, careful consumption, no waste, and greater harmony with nature.

A post-industrial house is designed to use renewable resources wherever possible in its construction; it is designed not to consume but to conserve — your time, your money, your paint, your comfort. It is planned to embrace the natural world rather than to ignore it; in that embrace it finds protection as well as warmth.

It is also designed (by you) to fulfill, to sustain, to cheer, to inspire, and to keep the rain off your head, the snow from your bed, and the wind from under your door. That's Charlie's part.

The multitude of fine illustrations, figures, diagrams, etc., were produced just for this book by Tom Paiement, a Maine artist, shelter builder, and Charlie's neighbor.

My part includes writing this introduction and a bit more, as you learn when you keep reading.

For more information about direct classroom and hands-on experience, you can reach Charlie at Cornerstones: The Wing School of Shelter Technology, 54 Cumberland Street, Brunswick, Me. 04011.

(JNC)

The Idea

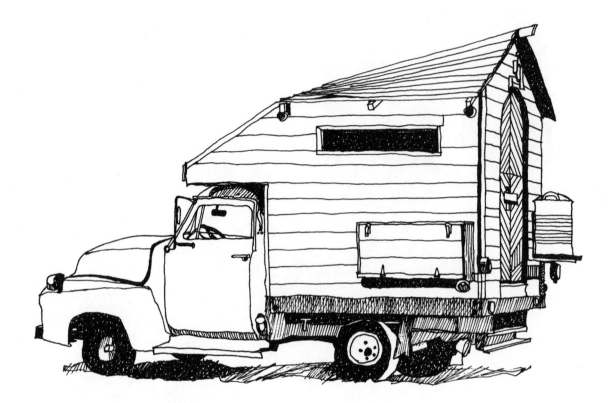

1

Why

Sailors, vagabonds, and wandering minstrels are houseless by choice. Apartment dwellers, owners of condominiums, and residents of mobile homes can live comfortably in mass-produced structures designed for average human beings.

Then there are those millions — like you, perhaps — who find themselves niggled by the notion, excited by the idea, and, finally, captured by the compulsion to build their own homes.

Like love, the origins of the notion are often a mystery. You, your husband, wife, lover, friend, children, relatives, and other fellow sojourners on your particular voyage through life may have lived happily for years in a third-floor apartment, in an old farmhouse (which you have restored), or in a perfectly reliable suburban split-level.

Then, the notion materializes. Who knows from whence it springs? But sketches of floor plans accumulate on the kitchen table; real estate sections in the local press are studied; and

hints are made to land-owning relatives. Dreams fade and materialize; hopes rise and fall with the tides of economic reality; visions of the home you now want become plans for the place you must have.

Your life has acquired a new presence, your days and nights are populated with a new breed of thought, and you begin to actually believe that the deed can be done — that you can plan, finance, and build your own shelter, your own home, your own place to live that will be exactly the way you always wanted it to be.

How innocently are cast the crucibles in which we are tested. Once captive of the conviction that you can have such a home, you must then decide who it is you are and what you want to be. This identity confrontation — which can become an identity crisis — could well be the reason why sailors and vagabonds have a reputation for carefree happiness, and why apartment dwellers and condominium owners can read the daily metropolitan headlines without evidence of shock to their systems.

Before the notion to build your own shelter becomes an exciting idea, and surely before that idea becomes a compulsion, ask yourself: Why? Why move from where you are? What is it you want that you do not now have? Does the place you think you want better serve you as a dream, or will its reality truly enhance your state of being? What is it you want the new shelter to say about you, about your particular two-, three-, four-, or more-person community? Where do you want to move to? What is the kind of landscape/seascape that most sustains you? Will your friends find you? Do you want them to?

As such inquiries arise and gain in stature, you should begin to realize that the decision to build your own home is also a decision to analyze and verify the details of your own existence and the existence of those who will share the place with you. Such an investigation is elemental. It requires a great deal of energy, honesty, talk, and time; it may well prove traumatic.

But it is essential. Getting your head in shape is the first and most important construction project in any home-building plan.

Advice from those who have done it can help you. Here is a little:

Try not to be in a hurry. It is difficult, once the enthusiasm is on you, to consciously apply any psychic brakes, but you should try. A self-planned, self-built shelter — even a modest one — is a mass of details joined over a period of hours, days, week, and months until they become a whole. None of the details will be quite in order until it is directed by an overall philosophy, guided by a masterful purpose. You must develop that philosophy, chart that purpose. Such achievements are generally not possible overnight. Take six months, take a year, or better yet, take two.

Some of the decisions will be intensely per-sonal. Do you want to sleep alone in a room, in a single bed, in a double? How far do you want to travel from the bedroom to relieve yourself? Do you want stairs, or no stairs? Can you climb a ladder? What, precisely, is the nature of the relationship with the person(s) you will share the shelter with?

There is an awesome intimacy to the sort of self-investigations that should be made if your shelter is going to serve you well. But these are probes that only you and your family can make; our only admonition is that you give them the time and consideration they must have.

Our purpose here is not some sort of structural psychoanalysis. We will not (and could not) tell you how to get your head in shape for building. We can say, unequivocally, that any shelter you plan, design, and build will work well in direct relationship to the intensity and honesty of the effort you spend on answering two questions: Why do you want your own place? What is the true identity and philosophy of the beings who will live in it together?

Such is the spiritual nature of the shelter.

Having asked those questions, and answered them after a fashion (which is the best any of us can do), you should know whether you have any interest in the sort of shelter we favor. You can learn from this volume, regardless of the kind of shelter you plan; but you will find that the information herein will guide, sustain, and give you continued confidence if you share some of the conceptual cornerstones of a post-industrial shelter.

Lined up like two cornerstones in a row, the major ones include:

• the belief that shelters constructed in sensitive harmony with nature will not only be more comfortable and efficient (warmer in win-ter and cooler in summer) but also more agree-able. You'll be happier just being there.

• the conviction that homes need not be

victimized by industrial-age pressures for mass production, mass consumption, and mass waste; but instead should be built with an eye to community production, careful consumption, and almost no waste.

Such is the post-industrial ethic.

Pursued, it can lead your shelter plans along many different paths. Avoiding mass consumption, for example, can mean avoiding the use of paint. There is very little need for paint. Almost any dwelling can do better without it. Paint also means repaint, and that means a life sentence of consumption for the homeowner. Those who have planned post-industrially find themselves doing quite well without paint.

Thinking in no-waste terms can lead you to reuse building materials that might otherwise be headed for the incinerator or the dump. Cobbles from a street ripped up for a superhighway can make a lovely hearth; old boards from barns and other weathered buildings that are threatened by bulldozers can make handsome walls; there is a home we know with a fine entrance hall floor of Carrera marble — the slabs came from the demolished men's room of an abandoned post office.

Once oriented to the possibilities of community production, you will learn that there are local sources of materials and labor that you might not otherwise have discovered. Shelters built during the summer, for example, can benefit significantly from the availability of student workers. Most high schoolers consider hammers, saws, and planes friendly and exciting tools; the students want to learn how to use them, they want to be doing something, they want to be earning some money. If you know what you want your shelter to be and can communicate, you'll find students most responsive listeners.

Sources of supply for local materials are just as responsive. Once the word gets around your neighborhood — whether city or country — that you need local lumber, you'll learn about markets whose existence you never imagined. A family we know needed some doors. The word went out; the doors arrived from a decaying summer hotel that was being refurbished. The doors did not fit the dimensions, but their hardware — knobs, hinges, and latches — was removed and is in use in all its nineteenth-century glory.

There is no end, of course, to the shapes in which an energetically maintained post-industrial philosophy can materialize. (Skylights have been made from discarded Budweiser signs — the plastic ones lit from the inside that you see hung from poles outside country stores.) The key to making the most of an idea is to keep reminding yourself of the cornerstones: community production, careful consumption, and almost no waste, along with the need for natural harmony.

"Living in harmony with nature." It's a kind of greeting card phrase, right up there with "To thine own self be true" and "No man is an island." That's what happens to verities. They get reduced to Valentine poems and Mother's Day remembrances.

Instead of being condensed to their shortest form, aphorisms of this kind serve better if they are extended to their limits, and beyond. The thought of coexistence with nature may seem vapid initially unless you recognize how little thought has been given to such coexistence in the past century. With the discovery of low-cost energy, with the development of technology, and with the fruits of mass production that energy and technology made possible, a great many shelter designers either ignored natural presences or challenged them. The result in both cases has been a painful dis-harmony.

Entire American cities are built on the premise that heat and cold can be ignored, that rain and snow can be disposed of as nothing more than nuisances, that the winds are irrelevant and rivers are meant to be the slaves of man.

Science, technology, and "know-how" make this conquest possible. If a plane crashes into a mountainside in a thundersquall, if an ocean liner sinks in a typhoon, if a city's reservoirs run dry, a technological failure is blamed; the forces of nature are either ignored or considered of marginal influence. There is always a specialist somewhere who can come up with a remedy. There is always the confidence that as he learns more about science, the specialist can — once and for all — overcome nature, cure cancer, feed the world, and build a house with a roof that never leaks and a basement forever dry.

During the first half of this century, most Americans absorbed this prevailing man/nature relationship by a kind of osmosis. Little thought was directed to the potential costs of the assumption that conquest, not coexistence, was the way to go. Most of us came to believe that specialists, and particularly scientific/technical specialists, are essential to the success of even the simplest and most basic endeavors of our lives. Our hair has to be cut, our teeth polished, and our children evaluated by specialists, just as our homes have to be built by architects, contractors, carpenters, plumbers, electricians, masons, glaziers, roofers, and backhoe operators.

The very presence of these specialists derives from the concept that nature can be conquered. It has become "un-natural" for a person to conceptualize about building his or her own shelter. Many persons are afraid to contemplate the prospect on their own. They cannot take the first step toward their shelter without thinking they need an architect, an engineer, and a contractor.

This is an industrial-age phenomenon; it has not occurred before in the history of humankind. Quite the opposite. Until the mid-nineteenth century it was historically natural for man to build his own shelter; and because that was so, those shelters sought to coexist with nature rather than defy it.

Thus there is a dual benefit that can accrue to the post-industrial shelter builder. By extending the concepts of natural harmony to every facet of planning and construction, it becomes the natural thing to plan, site, design, and build the place yourself. Having taken that step, you will also try to create a shelter that understands the sun and can benefit from it, that knows the winds and can share space with them, that treasures the seasons, comprehends the tides, is unsurprised by the frost, and knows there will be thunder.

In the process of thus coming to terms with nature, you may discover you are also restoring self. Although there will be times after you embark on your shelter voyage when you wish you had never set sail, once arrived in the place you make for yourselves, once certain that you know and understand the natural forces that affect it, once confident that you can not only build a shelter for yourselves, but can do it so it becomes a harmonious presence in your life and in its environment, then you will be privy to a measure of self-knowledge that can sustain you for the rest of your days.

That is a piece of spiritual furniture that you may discover after you move into the post-industrial shelter you have made for yourselves.

(JNC)

2

HOW

Most visitors to our home are there just a short while before they ask: "Who designed your house?"

There is an originality to the place, a difference, an out-of-the-ordinary-ness that makes the question a natural, and even courteous, one.

When I reply, "We did," the response is something less casual. "You mean you did it yourselves . . . you didn't have an architect . . . surely you must have had help . . ." and so forth.

I am repeatedly surprised at how exceptional the reality of an owner-designed home is to so many people. Before we got into the project (and this was our first), I had always thought that architects designed bridges, pyramids, and skyscrapers, with perhaps a museum thrown in now and then. I had spent forty-nine years with the assumption that most would-be new homeowners (unless they were buying a prefab or mass-produced house) designed the homes they built. My notion of an architect in such circumstances had much more to do with a fellow who handled the technical details (like where to

acquire the proper doorlatches) than it did with a person who laid his/her own creative plan on the would-be home builders.

We learned differently early on in our shelter saga. I discussed the project with a young (and excellent) architect, a friend of the family. Six weeks after the discussion, back came a lovely, three-dimensional model and a preliminary set of prints. They looked nothing at all like the rough sketches I had drawn. The place was a completely new presence, quite dramatic looking, but also quite alien.

The arrival of that model was an intense education for me in the realities of dealing with architects. I had much personal affection for the fellow we had encountered, so I thought it best to stop with the model, rather than ask him to go back and try again. The cardboard and papier-mâché version of the place I had never seen before cost me $2,500. Some lessons are more expensive than others.

But, as events evolved, my education proved a bargain. Once I had shelled out the hard cash for a cardboard minihouse that only the tiniest dolls could enjoy, I had also made a commitment to putting myself and the family into the design business.

We began with our spiritual values. I suggest you look rather closely and honestly at yours before you begin. The experience can be enlightening, painful, illuminating, and perhaps too revealing — particularly if you discover for the first time that your value systems are not as compatible as you had assumed (such assumptions being made on the basis of illusion rather than reality).

Fortunately, we agreed on the basics: a desire to design in harmony with nature; to build a place that conserved rather than consumed energy; to bring the outdoors inside as much as possible; to utilize simple structural principles that amateur carpenters could conquer; to give a feeling of security and coziness to sleeping spaces; to utilize secondhand (and more) building materials wherever possible; and to do the entire job so future maintenance would be cut to a minimum. We also agreed we wanted to be comfortable, we did not want to "suffer" on behalf of our conservation. We wanted privacy, a sense of isolation from the world and neighbors; and we wanted a place both we and the children could enjoy freely, with neither party expected or required to make too many concessions to the other.

We opted for wood wherever possible, for no linoleum anywhere, and no plasterboard. We hoped to grow old in the house. It should, then, be designed for easy cleaning and caring, preferably on a single floor. The children, we hoped, would grow older elsewhere; their rooms would be arranged flexibly, so they could perform other functions in the event that when the boys reach forty, they decide to strike out on their own.

Each of these decisions springs from a major

value basic: how you are affected by the world around you; how secure you feel with the open sky and sunlight, or snowstorm; how much of a materialist you are; how much of your compulsion to consume you can overcome; how you value "secondhand"; how much you love your children and your spouse.

Just who is everybody who will live in your shelter?

Heavy, the going gets very heavy.

Don't push it. The time it takes to design your place should be about twice or three times as long as the time required to build it.

If you complete the project, you will find your family community knit by sinews that will survive every other challenge.

At the beginning, you must stay loose. You must talk, draw, and discard, then begin again. I learned my own personal lesson from the process. Compulsive about getting done what needs to be done on the daily accomplishment list, it was agonizing for me to spend several high-energy hours drawing floor plans, only to have them tossed. But I learned — how I learned. Only I remember those early designs — plans that would keep me from sleeping with their excitement. Now, I am delighted they never materialized as our home.

If you can manage the value search and the looseness, then acquire the paraphernalia. It helps.

Even modest art supply stores have the drafting materials that can make the job precise and relatively easy. In my rummagings, I was surprised to find architects' stencils that had bathtubs and refrigerators cut to scale. Before that, I had measured every bit of basic furniture in our old home and had tried to reduce it accurately on a quarter-inch-to-one-foot scale. The stencils do it for you, and better.

We'd spend hours pondering the dimensions

of rooms we liked, and then try to convert those dimensions to the new overall plan. It always helped with difficult tussles to get back to the values.

We had wanted simplicity, a harmony with nature, the outdoors indoors. There is nothing simpler than a shed roof rectangle. We drove around, looking at all we could find — garages, henhouses, and standard homes. After pulling out my tape (indispensable for amateur designers) to measure an agreeable garage one afternoon (while the owner was absent), I decided that a shed about 24 feet wide and about 15 feet high on the high side with a three-inch pitch to the roof had a decent set of proportions. Rather classic, and simple.

That became our basic unit, with the high side facing south, admitting the best of the outdoors, and the north side short, keeping out winter to conserve energy. And the boys could build a shed if they could build anything.

I'm telling you this in such detail because I'm trying to describe the process. Once it becomes clear to you, your design will follow. But you cannot expect success with superficial preferences. You must get into what it is you truly want. If you'd rather not deal with that tender topic (and there are scores of good reasons for avoiding it), then forget about design and buy a prefab standard.

Once you have the basic form of the house you want, build a primitive (but scale) cardboard model. When we had opted for the fundamental shed, I put in rooms for us parents, plus rooms for seven kids, plus friends, plus baths, plus kitchen/eating, living, entrance, and storage, plus laundry, freezer, and dogs. With most rooms opening to the sunny south, and most as compact as possible, except for a large living room, I ended up with a shed 103 feet long by 24 feet wide. In model form, it looked like a boxcar that had been through a wringer.

But when I sliced across the shed in two places, and began moving the resulting three smaller sheds back and forth from a center line, I got some wonderful variations of space and plane.

That's what the place is: three sheds cut from one pattern and joined in what we think is an interesting variation.

It took most of a year to do that. We completed rough floor plans for the inside divisions, windows, doors, etc., but we never did get the inside finished. It changed too fast as we built the place because others who saw the house developing also created and added brilliant design features.

The boys decided to save space in their bedrooms by building bunks aloft, like Pullman berths. I wasn't around when they made that decision; as carpenters they acted on it before I arrived for my daily check on their work.

The plumber and the electrician had ideas, and good ones. An architect also helped. He came on as a consultant and contributor, however, and not as an overall honcho.

Another architect visited with a friend who had come to see how we were progressing. He suggested crossbeams on the upper windows to keep the house from racking, and in went the beams. An engineer suggested cable trusses on the long beams, and the cables went up.

I remembered the barns where those beams had come from, and put in two sleeping lofts that had never even been considered when the house was on paper.

The design went on every day for almost a year before the house was begun; it continued every day while the place was abuilding; and it has been continued in all manner of subtle ways since we moved in three years ago.

That's what can happen if you begin with the basics, if you search your collective souls, if you determine your value priorities.

If you don't design from such profound fundamentals, you may have much trouble with the next, and final, step — construction.

There is very little that's pleasant about digging post holes in blue clay, or (as I did) spending the three loveliest days of the summer excavating a place for the heating oil storage tank. There is no joy in hammering nails, especially when you must do it from a ladder or bent to the floor. Sawing, heaving, hauling, pounding, measuring, moving, carrying, cutting, planing, splitting, drilling, shoveling, and dropping a two-by-six on your instep will never come under the heading of pleasurable recreation.

Nor will the constant conflicts with suppliers of building materials — those remote untouchable bandits who promise delivery in June and fail to make good by August. Oh, the fierceness and the futility of the anger that must be fo-cused their way. How it depletes, how it haunts the hammering.

The headaches and the hard labor of construction would be unbearable (as far as I'm concerned) if it were not your own creation that's at stake.

That's the real reason why design with personal integrity is so important. If you know that you are busting your asses to build an object of your own creation, then you can suffer every indignity of construction. Each day, no matter how minute the advance, you are closer to a realization of your dreams.

There are few such glorious opportunities in a lifetime. This one begins with comprehension and design.

(JNC)

3

How—
Some Details

True, there are architects so called in this country, and I have heard of one at least possessed with the idea of making architectural ornaments have a core of truth, a necessity, and hence a beauty, as if it were a revelation to him: All very well perhaps from his point of view, but only a little better than the common dilettantism. A sentimental reformer in architecture, he began at the cornice; not at the foundation. It was only how to put a core of truth within the ornaments, that every sugar plum in fact might have an almond or caraway seed in it, — though I hold that almonds are most wholesome without the sugar, — and not how the inhabitant, the indweller, might build truly within and without, and let the ornaments take care of themselves. What reasonable man ever supposed that ornaments were something outward and in the skin merely, — that the tortoise got his spotted shell, or the shellfish its mother-o'-pearl tints by such a contract as the inhabitants of Broadway their Trinity Church? But a man has no more to do with the style of architecture of his house than a tortoise with that of his shell — What of architectural beauty I now see, I know has gradually grown from within outward, out of the necessities and character of the indweller, who is the only builder, — out of some unconscious truthfulness, and nobleness, without ever a thought for the appearance; and whatever additional beauty of this kind is destined to be produced will be preceded by a like unconscious beauty of life.

— THOREAU, *Walden*

No two people will design a house in the same way. Design is not the sort of thing that can be successfully programmed. There is too much inspiration involved, and inspiration is a mysterious thing not subject to analysis. However, shelter design is *a synthesis of an incredible number of details.* I give here a number of design tools that may prove useful in the synthesis. They are given in what seems to me to be a logical order.

They include: program requirements, activity space analysis, time space analysis, circulation analysis, efficiency analysis, visual patterns, and architectural drawings.

Program Requirements

First, we must have a function list for our house. Written down, most usefully in order of priority, this becomes what architects term "the program." It is a shopping list to constantly remind us of our design goals.

Sometimes the program cannot be fully met. After all, one of the items in the program will probably be total cost. Most likely the program will be changed several times as it becomes apparent that some of the goals are contradictory.

Often a couple will ask me to arbitrate a conflict they are having over design goals. I refuse. Compromise is the death of exciting design. Excessive compromise in a marriage results in the couple becoming less than the sum of its parts. A good marriage will tolerate, if not nurture, the individuality of its members. A good house will express the individuality of its occupants. Rather than compromise, let each occupant design his own personal space, and then put the spaces together.

Space

Thoreau implies that the form of a house is ideally an expression of the occupants' life-style, just as the tortoise shell is an expression of the way in which the turtle lives. The house is thus a tool, an aid, in achieving a life-style, hopefully evolved by the occupants through self-knowledge. And what is life-style but activity? — working, making love, eating, and reflecting. Our lives are multifaceted and multispeed and are contained within an activity/passivity spectrum.

Activity takes place within, and can thus be described in dimensions of space and time.

Activity Space Analysis

Activity space may be divided into two basic areas as shown in Illustration 3. The *endospace* reflects a need for privacy and being with ourselves. *Ectospace* is social space, space for being with others. Just as a personality must graciously encompass both aspects, our shelter must harmoniously incorporate both spaces. A *meso* or transition space is needed to accomplish a gracious transition. Without it, a house or person would present a conflicting, awkward personality in which opposites continually clashed. Within the broad *ecto* and *endo* spaces, gradations will exist. For example, the *ecto* space itself may, at one extreme, be exhibitionist and flamboyant (a garden party space) while at the other extreme sociable but intimate (dining room).

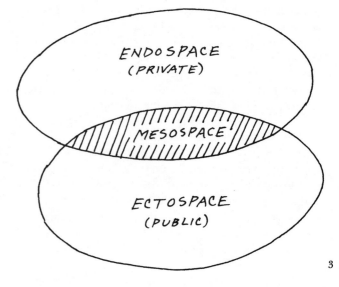

3

Time Space Analysis

Time can only be measured as a sequence of events. In life, time is measured in cycles: the one full cycle of birth to death, the annual cycle of seasons, and the daily cycle of the sun and our activity. Our lives are given continuity and harmony by a synchronism sensed between these cycles. The circle of Illustration 4 represents the three synchronous cycles: life, year, day. The driving force of all three cycles is the

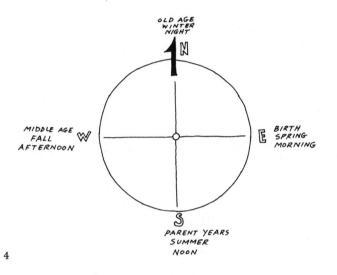

4

sun. It is the giver of life, the creator of the seasons, and the clock of the day. By considering the circle a compass, the dimensions of space and time are drawn together.

ACTIVITY DEFINED IN BOTH SPACE AND TIME

The shelter consists of enclosed spaces, defined in terms of activity. We are now in a position to graphically locate these activity spaces in terms of both space and time coordinates. Illustration 6 shows how one person might define his life-

style and its physical expression, the shelter, in both space and time. First, he has located his activities in the *endo-ecto* space dimension of Illustration 5. These defined activity spaces are juxtaposed in the shelter in such a way as

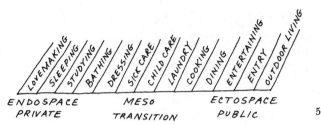

5

to preserve their linear space relationship. Second, a time association is given to all of the activities. The activity spaces are then related to the time/compass circle of Illustration 4. In the shelter, the activity spaces are thus related to points of the compass. We then arrive at the two-dimensional chart of Illustration 6.

In a real house, the space dimension corresponds not so much to feet and inches of physical separation as to psychological separation, achievable by horizontal distance, vertical distance, walls, or constricted paths. The most private space may actually be a loft over a relatively public space, the privacy being derived from visual separation and difficulty of climbing.

Circulation Analysis

The way people move around a house is known as the traffic pattern or circulation. Why should

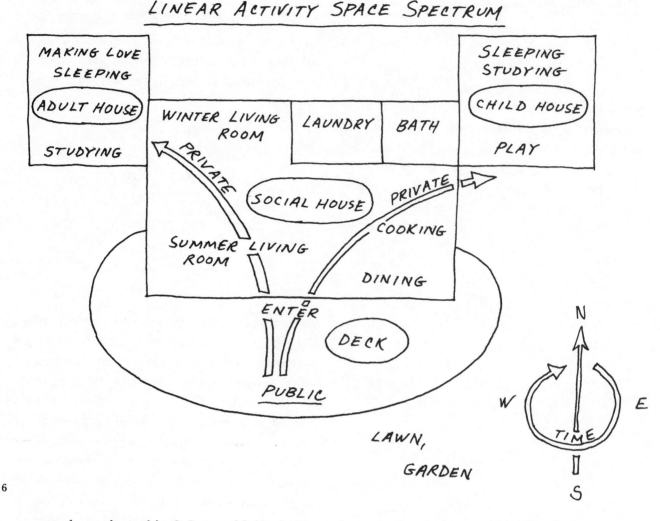

LINEAR ACTIVITY SPACE SPECTRUM

6

one study such a thing? I can think of two reasons: 1) We walk around the house a surprising amount and our walks will be more pleasant if the routes are enjoyable; and 2) Traffic lanes can be as disruptive and divisive in a house as freeways in a city. Traffic should be detoured around, or better yet, never have occasion to walk through, private or intimate spaces.

Illustration 7 is a tool that can be used in rating the success of a traffic pattern. Each major house area is assigned a circle. Arranging the circles approximately as they appear in the house is not necessary but will make the diagram more meaningful. Draw lines between all possible pairs of circles, representing trips between areas. The simple number of trips between areas

is not a true measure of traffic since no allowance is made for either length or difficulty of a trip. A long trip past a planter may actually prove less wearing than a short trip involving stairs. Therefore each trip is assigned a degree of difficulty. The overall success of the traffic pattern is judged by the total of trips multiplied by their appropriate degrees of difficulty.

Certain areas are easily disrupted by traffic. Obviously a family bathroom should not be accessible only through a bedroom. Entertaining in the living room is made more difficult when children must pass through in order to brush their teeth. Private spaces at the extreme left in the space spectrum of Illustration 5 should always constitute traffic dead ends.

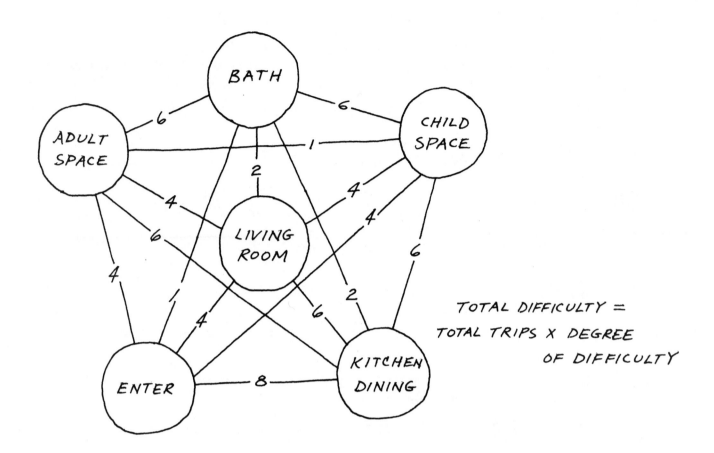

TOTAL DIFFICULTY =
TOTAL TRIPS X DEGREE
OF DIFFICULTY

CIRCULATION CHART

Efficiency Analysis

If a house cost $30 per square foot of enclosed area, the homeowner will have paid $2,400 for the privilege of entering bedrooms through a 5′ × 16′ hallway. He also will have paid at least a thousand dollars for the empty space within interior walls. Halls, walls, stairways, heating plants, and closets all represent areas of potential wasted space. Anyone interested in the efficient use of space would do well to study the interiors of sailing yachts, where even curved surfaces are used for storage.

For the life of me, I can't understand why kitchens must be camouflaged as French drawing rooms. Personally, I find the sight of

food and dishes on shelves exciting. With all of the food and utensils hidden away in miniature Mediterranean fruitwood closets, it's a wonder that all of the elements of an average meal are ever assembled in time. A thousand dollars could be saved, not to mention years of the future occupants' time, by modeling our kitchens after the commercial variety instead of furniture showrooms.

The machine room of a house serves the same function as the boiler room of a ship. What ship's captain would tolerate a boiler room where all of the plumbing, valves, and controls were inaccessible? Yet most of the plumbing and wiring in the average house is totally inaccessible! When there is trouble, the plumbing bill is accompanied by a carpentry bill. In my house all machines that whir, whine, or pump are located in the machine room, along with most of the plumbing and wiring. The wall spaces not blocked by machinery are covered with open shelves from floor to ceiling. Trouble with the plumbing typically requires removal of a few boxes of Corn Flakes — not the entire wall.

Visual Patterns

Philosophically I agree with Thoreau that the house should be developed from the inside, and hang the outside ornamentation. Practically, however, there are many exterior aspects of the house that may be varied without compromising the inside space. There is no question that the human eye generally finds some geometric patterns more pleasing than others. Most of these are in some way patterns found in nature, and by choosing them we only make our houses more natural.

The point Thoreau neglected to develop was that the pleasing patterns of nature have evolved over millions of years. In architecture, a similar process occurs. The classic styles of saltbox, colonial, cape, etc., evolved through a series of refinements. Their success was assured by a successive emulation of strong points and improvement upon the bad points until each style was optimized for the conditions of time and place.

The greatest single downfall of the owner/designer/builder in my experience has been the external appearance of the house. I recommend that you take time to critically study the exteriors of houses that please and displease you.

Try to identify the specific features that characterize the appearance of each. Study especially the ones most similar in general construction to your own conception. As an aid toward recognition of specific patterns, I list here seven architecturally recognized principles.

NATURAL LINES

A house will appear more natural in its surroundings when its major lines are extensions of the predominant lines of the surround.

PLAINS HOUSE

MOUNTAIN & FOREST HOUSES   8

RECTANGLES

The Greeks recognized a certain rectangle ("the golden section") as being the geometric

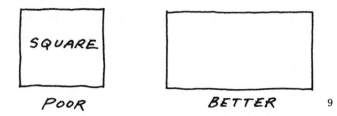

SQUARE

POOR   BETTER   9

figure most pleasing to the human eye. The exact proportions are not as critical as the simple preference over the square.

STABILITY

The eye prefers a visual illusion of stability.

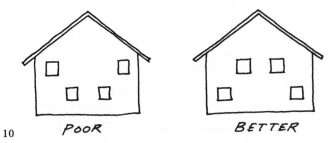

10   *POOR*   *BETTER*

HORIZONTAL ALIGNMENT

Place all windows and doors so that their tops fall along one line.

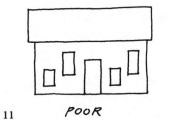

11   *POOR*   *BETTER*

VERTICAL ALIGNMENT

Place windows directly over one another.

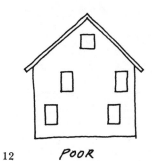

12   *POOR*   *BETTER*

SYMMETRY

A symmetric house benefits from symmetry, BUT an asymmetric house benefits from asymmetry.

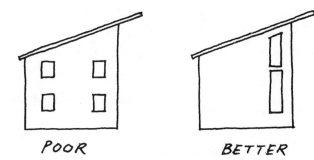

POOR   *BETTER*   13

SIMPLICITY

Avoid busyness or too many themes.

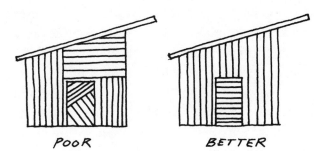

POOR   *BETTER*

BETTER   14

Architectural Drawings

I am dismayed at how few owner/builders recognize the importance of architectural working drawings. They apparently feel that architects draw because they like to play artist. While architects do take pride in a certain artistic value of their drawings, this artistic contribution is not one part in a hundred of the true value. I can sympathize with this feeling. When I first considered building my own house, I asked a local successful builder his opinion of drawings. "Never use them," he said. "The only time I did was the only time I made a mistake. The G.D. architect made the mistake!"

This attitude is well afforded by a master builder. Without knowing it, he has all of his plans in his head. Just ask him about some detail of construction and he'll sketch it out with dimensions to 1/16 of an inch without ever thinking. He knows his plans are right because he's built that detail a hundred times!

You, however, have never built a house. You have two choices: Build a practice house and make all of your mistakes before you start your own; or "build it" on paper with a sharp pencil and architect's scale. I recommend the latter. If the house doesn't work out all you've wasted is a piece of paper and a few evenings of fun.

Here is what a set of drawings, done carefully with a minimum number of drafting tools, will do for you:

(1) As you develop the details of the drawings, you will be forced to make design decisions that you otherwise could put off until actually confronted with the physical reality. A scale drawing would have shown an acquaintance of mine that his window glass was larger than the window frame opening and that the windowsill wasn't wide enough to extend beyond the wall sheathing, with the result that rainwater flowed into the wall space.

(2) Because you are actually building the house on paper, fastening problems become immediately apparent. How will the Sheetrock be nailed at the room corners? How will the wiring be concealed?

(3) Your material list will be precise because the number and dimensions of all framing members are easily calculated from the drawing.

(4) Angle cuts such as are required for rafters can be measured on paper instead of by intricate trigonometric calculations or by seat-of-the-pants guesswork.

(5) You will better know how the house will finally look on its site.

(6) You will have an efficient means of communication with:

    (a) the banker, who will be more willing to loan money. A picture may be worth a thousand words but it's also worth several thousand dollars.

    (b) subcontractors, who will dare to give tighter bids if they know precisely what is involved.

    (c) your workers, who won't require your presence at every turn. The amount of time your workers will not be sitting under a tree drinking beer while you fret over an unexpected problem will alone pay for your initial effort.

    (d) yourself. When everything seems to be going wrong and three people are bugging you over details any fool should understand, your blueprints will remain unperturbed.

(7) Last and most important, drafting will teach you a self-discipline that is required in efficient and low-cost housebuilding.

In summary, a month of evenings at the drafting board will save weeks of labor, as much as 25 percent of the cost of your house, and prevent the best experience of your life from turning into a nightmare.

Required Drawings

I cannot in this short space teach you drafting. In another sense no one can teach the skill. You must obtain an architectural drafting text or some advice from an architect and start drawing. The number of required tools is minimal. I recommend:

good quality 18″ × 24″ drafting vellum
an architect's scale
a parallel rule or T-square
a protractor or adjustable triangle
a fixed triangle, if you don't have the adjustable kind
a residential template
drawing pencils or lead holder
a good pencil sharpener

A complete set of drawings includes the following:

TITLE PAGE AND INDEX

This is used primarily to impress the bank board. If you want to exercise your artistic talent, an artist's rendering or perspective drawing will dress up this page.

SITE PLAN

Perhaps the most important of all, this plan shows the original contour of the land, proposed finished contours, original trees, trees to be left standing, well, septic tank, driveway, electrical service, placement of house, and orientation. Other important information such as direction to best view, prevailing winter and summer winds, shelter belts, etc., may be included. This is the most important drawing because it is an aid to clear thinking regarding the most important aspect of your shelter — the site. Don't sketch the contours as you imagine them to be! Rent a surveyor's level or buy a "pop" level (a five-to-ten-dollar device), and some fine weekend measure the contours on a 1-foot-height interval for a distance of at least 100 feet around the proposed house site. In drawing the site plan you will need a reference direction. This may be a line between two permanent trees or the direction toward a landmark from a pipe in the ground. Whatever it is, make sure it will remain there until the foundation is finished.

FOUNDATION PLAN

After generating the site plan, the foundation plan is simple. The height of sill above grade and amount of excavation and grading are simply computed. This plan is important because foundations are messy and discouraging. Unfortunately, the psychological low point in house construction usually comes during construction of the foundation. All too many people give in to the mud and call a contractor at this point, not realizing that the psychological high point comes the very next day with the placement of the first floor joist. At this point, having a good foundation plan will keep your spirits afloat.

FLOOR PLAN

The floor plan shows thickness of walls, location of all doors and windows, major appliances, and the real dimensions of rooms. Additional copies of the basic floor plan are used to show furniture placement, wiring circuits, plumbing runs, and heating apparatus. Practiced architects sometimes manage to show all of the above in one floor plan, but my drawings get too cluttered. The floor plan is the basic tool in sizing rooms and judging circulation. It therefore is the one most often redrawn.

ELEVATIONS

An elevation is a straight-on drawing of the lines of a house from its four or more sides. You may settle for a front and one side view. These are used primarily in creating the visual patterns of the house exterior. Thoreau said we don't need it, but I never reject anything useful, provided it doesn't compromise the more important interior values. At least, with elevations, you won't be totally surprised by the appearance of your creation.

SECTIONS

The *framing section* shows foundation, sills, joists, posts, and rafters and is used in computing mechanical loads, strengths, and moments. It is also useful in assembling the framing materials list.

The *typical section* is a cutaway view of all of the material used in foundation, floor, wall, and roof. It is drawn as if a giant knife were sliced vertically through the most typical part of the house. Because it shows so much detail, it is usually drawn at a larger scale than the other drawings.

DETAILS

One or more sheets may be devoted to large-scale drawings of detailed features, such as the mullion of the homemade double-glazed window framing or the kitchen cabinets.

HEATING, PLUMBING, AND ELECTRICAL

If not shown on the floor plan, another copy of the simplest version of the floor plan is used to indicate the location of pipes, valves, cleanouts, and vents of the plumbing system; circuits, fixtures, and boxes in the electrical system; and heating sources and plumbing or ductwork of the heating system.

But before you start drawing, you need some land.

(CW)

The Land

4

Where

There is, according to novelist Lawrence Durrell, a "home landscape" locked within every soul. It is a geographic specific, a bit of land and/or water that is at the memory base — the place you return to in your deepest dreams, the place that attends your subconscious; it is the nucleus of your memories and at the center of the mind's-eye scenery that appears when you visualize your "home."

Most of us can readily identify such a location, but only some of us may be convinced that it is also the place where our house will be built. What was a marsh in our youth may have become a shopping center; the prairie can disappear just as the mountains can be mined and rivers dammed. The mind's eye sees the past; the present and future can be quite different.

Assuming that you are as free as a plover, able to move from Lapland to Tierra del Fuego across thousands and thousands of miles, where in the world would you want to live?

Most of us cannot make that choice. We have jobs, families, roots, obligations, neighborhoods, security, identity, and inertia working to keep us where we are. But it might be well to consider the plover's purposes, assuming you are as free as a bird.

Putting aside the still unresolved migrational mysteries of the eons, it appears the plover travels with the sun and the seasons. It is the ranges of temperature, the prevalence of the winds, and the length of the days that the plover follows from one end of the earth to the other. Being free and eminently mobile, the bird moves with the climate conditions it likes best. Without shelter, the plover finds its home within a framework of moving weather standards, even though it must spend much of its life aloft to keep up.

More pedestrian creatures, like the bear, endure the climate they have, avoiding some of winter's discomforts by finding a protected den and sleeping from October to March.

Between the plover's constant movement and the bear's half-year of immobility there is the entire natural sweep of climatic adaptations made by every creature — except humankind, and, more specifically, the inhabitants of the industrialized nations.

Since the advent of science and technology, the human animal has found conquest of climate more appealing than adaptation. High-energy air conditioners can "conquer" heat and humidity; central heating systems can overcome cold; exhaust and intake fans take the place of wind; paintings of trees are hung to bring a reminder of the forest indoors.

Such mechanisms should be set aside when you begin to consider the general location — north, south, east, or west — of your post-industrial home. Rather than thinking like twentieth-century humans, you might more wisely and productively emulate the plover and the bear. That is, investigate the primal climate that most appeals to you. If the world is indeed your starting point, then reach into yourselves and investigate your relationship with the sea, your need for seasonal change, your comprehension of the cold, and your tolerance for persistent winds. From such investigations will come the clues that will lead to the natural locations that are most compatible.

However, if your starting point for a home site is geographically limited to a region, a state, a county, or (more likely) a town or neighborhood, do not despair. You, too, can learn from the plover and the bear. Their lesson is the relative compatibility of any climate once it is understood as an entity to be lived within, rather than overcome.

You may despair of the winters in your particular place more because you have never tried to understand them than because of any particular intensity they possess. The winds may taunt you only because you have never deciphered their secrets, and the rains may depress because you find only dampness in a downpour. Such are the legacies of technology.

They have prevented many of us from recognizing our essential naturalness, our climatic adaptability, and, in turn, the generally friendly nature of the climates around us. Remember the plover and the bear when you choose your general location. Be aware that there is more to be gained by adapting to nature than by shutting her out. Analyze the natural climate of the locations available to you, and recall that you are designed to live within that natural climate, no matter how adamantly you may have been persuaded otherwise.

If you doubt that generality, then we suggest you study the following specifics.

(JNC)

5

Where—
Some Details

Comfort Zones

out artificial aids over a range of 45° – 85° F!
Above this range artificial air-conditioning must

The human comfort zone is defined as the range of temperature and humidity over which the average human can perform sedentary tasks at maximum efficiency without having to resort to heavy clothing or artificial aids such as heat or air-conditioning. The zone is somewhat individualistic, depending not only on the individual but the culture. Illustration 16 shows such a comfort zone, which applies, on the average, to a person in the moderate climatic zone of the United States. The enclosed area is the area of comfort with *no wind* and *no sun*. The curves above the comfort zone show how the upper limit of the zone may be raised by *air movement* over the body at various rates. The curves below the comfort zone show how heat *radiation* on the body (sun or stove) can lower the minimum temperature of comfort. Note that both of these aids are naturally available to the house designer. In other words, the average person could feel comfortable with-

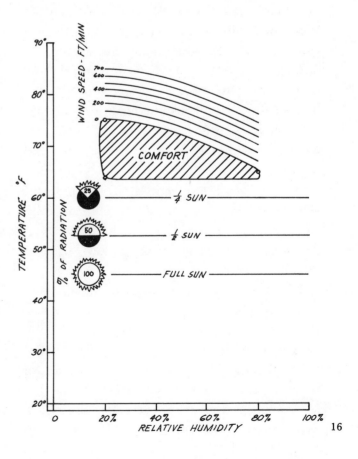

be used. Below the range heat must be supplied.

Chickens have yet to stop laying eggs just because the temperature dips to 0° F. They might appreciate a little heat but they don't seem to suffer. Dogs spend evenings under the wood stove, where it is at least 90° F, but also spend all day playing outdoors at 30° F. The deer, seen in the field all summer, herd up in secluded thickets, protected from the wind but open to the sun in the winter. Cows are found on the north side of a hill on hot summer days. But human beings are found in little cubicles within which the temperature is controlled to about 70° F by automated machines (furnaces) so complicated that only trained professionals can tamper with them.

All of nature's creatures (we seem to have forgotten that we are included) have evolved in a seasonally variable climate and have, therefore, acquired an adaptability to a yearly range of climatic conditions. We may wish summers were cooler and winters warmer, but is it not a fact that 50° F seems balmy in January but positively frigid in July? The comfort zone reflects this by moving up and down a few degrees seasonally. How else could the standard house temperature change from 72° F ± 1° F to 68° F ± 1° F within a few years without the entire population expiring?

Studies have shown that laborers who spend eight daytime hours at 30° F and the other sixteen hours at 65° F inside have *not more but fewer* colds than office workers at 72° F around the clock.

We all recognize the beneficial effects of a moderate amount of physical exertion. We have evolved with organs that expect, and have come to depend upon, a little stress and strain. Similarly, climatic stress and strain represents a form of exercise from which our bodies benefit. Just as running a marathon every day is a little extreme, we don't have to suffer alternately

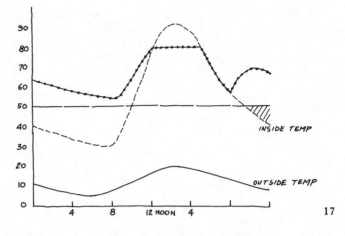

from frostbite and heat stroke, but neither do we have to be enslaved by the thermostat.

My house does not have central heating, nor does it have a thermostat in every room. It has built-in climatic variability. We will understand this after discussing the subject of house heating in Chapter 19. Illustration 17 shows the temperatures outside and inside my passive solar house. A passive solar house is one having large amounts of glass on the south side and small amounts of glass, or none at all, on the north. My house has 600 square feet of glass on the south, southeast, and southwest, and only 8 square feet on the north.

In the graph the outside temperature (solid line) reaches a daily high of 20° F at around 2 P.M. and a low of 5° F at 6 A.M. The dashed line represents the temperature curve this passive solar house has followed without any human intervention (including no source of heat). Starting from a low temperature near freezing, the house warms rapidly upon sunrise. The maximum inside temperature occurs at about the same time as the outside maximum, but the ΔT (temperature inside minus temperature outside) is 70° F. As the sun falls so does the temperature, with a rapid rate of fall starting at sunset. Through the night the rate of fall decreases, being proportional to ΔT. The inside temperature never reaches the minimum outside temperature during the sixteen-hour dark period. Notice that the average inside temperature is considerably higher than the average outside temperature. This is due to the greenhouse effect.

Clearly, however, this inside temperature range exceeds anyone's comfort zone. The normal temperature curve the house follows with my intervention is shown as the dotted line: As soon as the temperature inside reaches my upper limit of 80° F, I open a ventilator allowing the excess heat to escape. When the temperature begins to fall noticeably (around 4 P.M. during winter) the ventilator is closed. At sunset a fire is started in the Jøtul 118 woodstove. A quick burst of heat (producing lots of red-hot coals) quickly brings the temperature up to 70° F for the evening social hours. At bedtime the stove is restoked and the damper set proportionally to the expected overnight low temperature. This causes a constant rate of heat production and the house slowly cools down toward a morning minimum of 50° – 60° F, following the outside temperature. The Jøtul 118 is capable of burning 10 to 12 hours at a low setting, and so in the morning we have a choice of letting it burn out (sunny day) or restoking it for another 10 to 12 hours (cloudy day). We operate this way, using about four cords of wood per heating season (7,700 degree days) to heat 1,700 square feet of house.

If I found wood hard to come by, I could install reflective window shades or insulating shutters and reduce the fuel consumption to two and a half cords.

The temperatures in Illustration 17 are typical of the main living area of the house — the "social house." I earlier referred to the climatic variation within the house. Not only does temperature change with time, but with space, also. The adult house has two spaces — a full-sized loft that, without ventilation, is always 10 F° warmer than the social house, and a downstairs workspace, 10 F° cooler than the social house. In the early morning the cat and I are found in this loft. By late morning we find each other in the social house, and at 2 P.M. on a sunny day we work (I work, she dozes) in the cool workspace.

We, as a culture, have achieved, in spite of the Walt Disney hour viewed electronically as a series of TV lines and dots in "living color," an all-time low sensitivity to nature as a dynamic environment. Man created shelter as a buffer against the extremes of weather. It is not until recent times that we have completely isolated ourselves from climate.

However much we try to ignore it, weather is in our blood. Talk about weather cannot be entirely dismissed as an avoidance of intimacy. It is important to us. I used to work in a windowless basement office. All visitors from the outside world were greeted by, "What's the weather outside?"

I have been reborn. With the exit of the thermostat came a rebirth of awareness. I no longer call the weatherman; I don't need to. Like the farmer, I have developed an early warning system for subtle changes that foretell the future. I don't need to consult the almanac for the times of sunrise and sunset; I am there. Sunrise is when the beneficent rays begin to warm me; sunset is when the fire is lit. I don't need a clock; I am a sundial. I don't need a thermometer; the outside temperature is obvious. I don't need a calendar with the date etched by the angle of the noon rays against the wall. I know when the maple sap flows; it is when I should stop cutting firewood because the wood is at its greenest. I accept and exult in my nature.

Climate

Most people recognize the importance of the foundation of a house — if the foundation fails, the house may also. But whether the house works, is successful, depends upon something that too few consider — the proper choice and utilization of the site. Climate consists of two parts. The macro and the micro. *Macroclimate* is defined by weather bureau statistics: maximum and minimum temperatures, number of degree days per heating season, percentage of sunshine hours, wind velocity and direction, etc. These figures depend primarily upon geographic latitude, elevation, and proximity to large bodies of water. *Microclimate* expresses the local modification of macroclimate by features peculiar to a site — site slope, orientation of slope to sun, type of vegetation, directions of winter wind and summer breezes. Everyone is aware of macroclimate, as witnessed by seasonal migrations of affluent people to the south in winter and the north and coast in summer. In nature, the parallel is found in bird migration. But very few people are aware of the degree of microclimatic variability in their own locale. The ancients were aware of microclimate. The very word "climate" derives from the Greek for "slope" — a recognition of the extreme effect of slope and its orientation to sun.

Macroclimate

TEMPERATURES

Heating and ventilating engineers use tables and charts of macroclimatic data to determine: 1) the total amount of fuel required to heat a building for the heating season, and 2) the size of the heating equipment (heat produced per hour) required to maintain the desired inside temperature on the coldest day. This information, gathered and analyzed by the weather bureau for nearly a century, is actually more valuable to us than to the heating engineer. We, like sailors of old, can navigate our responsive craft through the weather, stoking the fires in heavy weather and scudding joyfully through solar days. The engineers build submarines — isolated, unnatural environments — plying the depths by brute force, oblivious of the weather.

We can use our macroclimatic data for gross determination of: 1) the natural seasonal climatic variability for our region; 2) the availability of the natural comfort aids of solar radiation and ventilating breezes; and 3) the extent of artificial comfort conditioning required by our shelter/site combination.

Illustration 18 shows *summer dry bulb design temperatures* (simply the reading of the ordinary thermometer). This number is used by the ventilating engineer to specify the capacity of air conditioner needed to maintain a cubicle at X degrees when the outside air is that temperature. Naturally, no account is taken of the local modification of outside temperature by microclimatic effects since: 1) the effects are not known by the engineer, and 2) he is probably safe in assuming that the contractor has either removed all vegetation likely to produce relief, or chosen a perfectly flat site with minimal vegetation.

Illustration 19 shows *winter design temperature,* the corresponding number used by the heating engineer to size the heating machinery to overcome the minimum annual temperature.

Illustration 20 is a map showing heating season degree days. The number of *degree days* (DD) in a day is defined as: DD = (65 – average temperature for the day.)

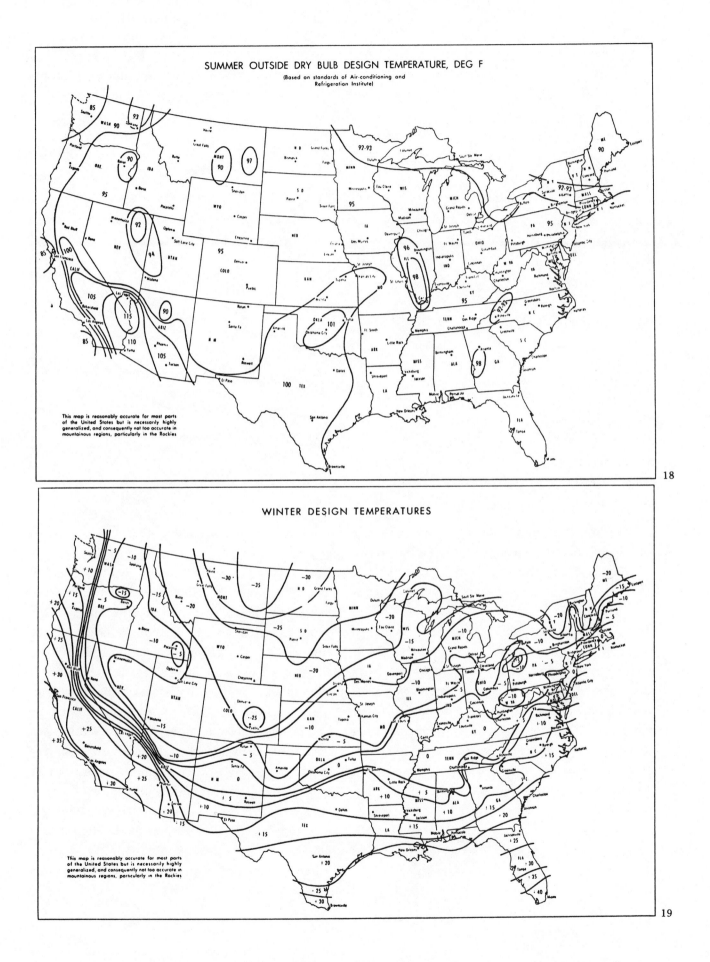

SUMMER OUTSIDE DRY BULB DESIGN TEMPERATURE, DEG F
(Based on standards of Air-conditioning and Refrigeration Institute)

This map is reasonably accurate for most parts of the United States but is necessarily highly generalized, and consequently not too accurate in mountainous regions, particularly in the Rockies

18

WINTER DESIGN TEMPERATURES

This map is reasonably accurate for most parts of the United States but is necessarily highly generalized, and consequently not too accurate in mountainous regions, particularly in the Rockies

19

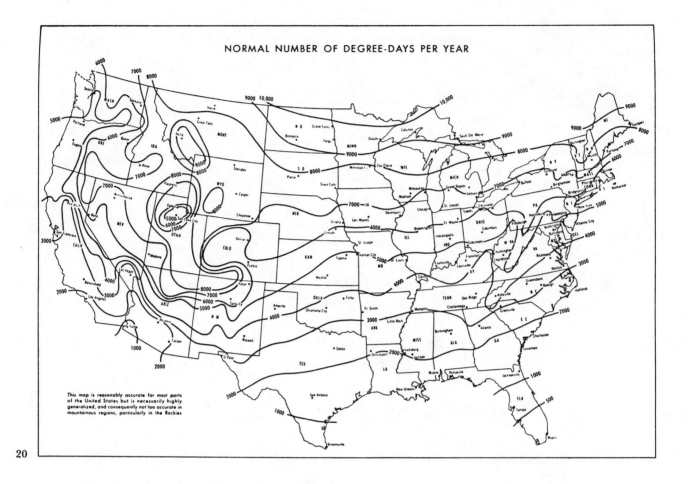

NORMAL NUMBER OF DEGREE-DAYS PER YEAR

This map is reasonably accurate for most parts of the United States but is necessarily highly generalized, and consequently not too accurate in mountainous regions, particularly in the Rockies

20

The number of degree days in a year is simply the accumulated number of daily degree days. Degree days are a pretty good measure of the amount of energy consumed in maintaining a constant house temperature of around 70° F. The difference between 70° and 65° F is due to the contribution to heating of sunshine, cooking, electric lights, and human activity. Note that the degree-day formula assumes maintenance of a constant indoor temperature of about 70° F. Some people are now maintaining their homes at 60° F as an economy measure. If your personal comfort zone allows that, you'll thereby save approximately ten DD per day during the coldest months.

SOLAR RADIATION

The most powerful natural winter comfort aid is the sun. But improper use of this tool can cause great discomfort at other times of the year. In order to predict the behavior of the sun, or the interplay between the sun and our shelter, we must be able to predict the path of the sun and the intensity of its radiation.

The path of the sun (the sun's position in the sky at any date and hour) is described by two angles: *azimuth* and *altitude*. Azimuth is simply the angle through which you must turn, starting from either north or south, in order to face the sun. Conventionally, one starts from whichever direction, north or south, results in the smallest angle.

Altitude is the angle above the horizon and

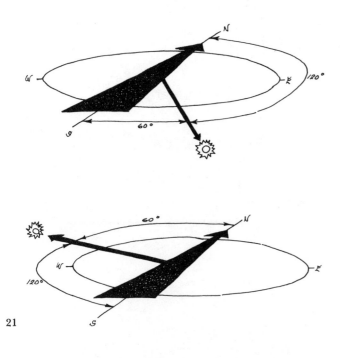

21

never exceeds 90 degrees. The altitude is zero when the sun rises and sets over the ocean or a flat plain and 90 degrees if directly overhead.

The solar angle charts of Illustration 23 give the sun's position in azimuth and altitude at all times of the year. To get some notion of what the sun can do for you, find the chart nearest to your site in latitude. After finding the sun's position on charts of greater and lesser latitude than your site, you can interpolate for exact angles. The charts consist of two superimposed patterns. I call them oranges and canoes. The orange (circles and straight lines) is a map of the sky with the distance of a circle from the outside corresponding to *altitude* and the straight line pointing toward *azimuth*. That is, a point at the very bottom perimeter of the chart represents a point in the sky which is due south and on the horizon. The point at the very center of the figure represents straight overhead. If you still have trouble envisioning this sky chart, imagine yourself lying flat on your back with your feet toward south and your head toward north. Now the chart is exactly what you see looking straight up. The

intersection of a straight line and a circle determine a single point in the sky described by an altitude and an azimuth.

The canoe is a series of sun paths across the sky. The lines (canoe ribs) running across the paths show the hourly progression along the paths through the day in standard time. Notice that there are only seven canoe lines. The seven lines are the precise paths for the twenty-first day of each month, identified by Roman numerals. Since the sun traces the identical path twice each year (March 21 and September 21, April 21 and August 21, etc.) we need only seven lines. For dates other than the twenty-first, interpolate between the two appropriate curves.

Charts as full of information as these are best understood by example.

Example 1:
For a site of latitude 44° N, where is the sun at 3 P.M. standard time on November 21?
  Month is XI (canoe XI).
  Time is 3 (rib 3).
  The sun rises in the east (at the right side of the diagram) and follows the curve labeled XI toward the west. At 3 P.M. the sun is exactly at the intersection of the two curves, XI and 3:
  Altitude (distance from perimeter toward center) = 13°;
  Azimuth (read at perimeter) = S 44° W

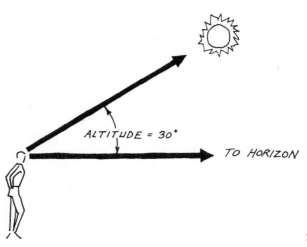

22

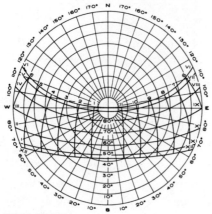

24° N LATITUDE

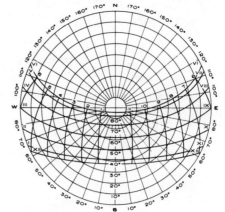

28° N LATITUDE

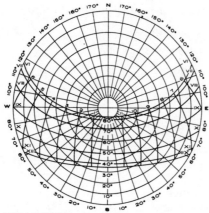

32° N LATITUDE

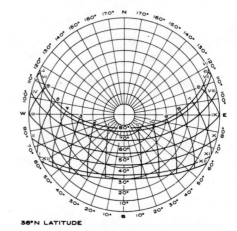

36° N LATITUDE

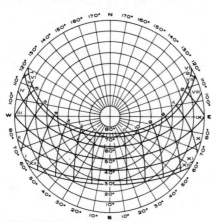

40° N LATITUDE

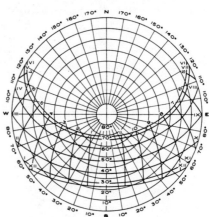

44° N LATITUDE

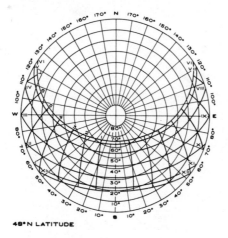

48° N LATITUDE

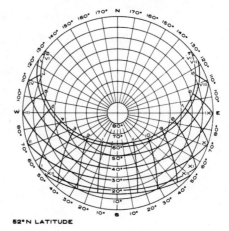

52° N LATITUDE

23

If I went to my house site at 3 P.M. on November 21, turned 44 degrees toward the west from south, then raised my eyes 13 degrees, I would be looking directly at the sun. If a year later, I looked out my south-facing window and did the same thing, there would be the good old sun.

Example 2:
For a latitude of 40° N, where and at what time does the sun rise on July 21?

  Month is VII

  Time is?

  Altitude = 0 degrees

  Looking on the 40° N Lat chart we find that canoe VII starts at altitude 0 degrees (perimeter) at just before 5 A.M. and at an azimuth of about S 115° E. That is, it rises, not due east, but 25 degrees to the north of east at 5 A.M.

Suppose that we build our house, as in Illustration 24, with window A placed to catch the earliest morning's rays. We will be disappointed in the summer

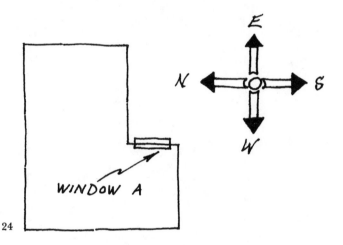

WINDOW A

24

because the sun won't fully illuminate window A until 7:30 A.M. when it comes around to due east.

Example 3:
You require direct sunlight in the morning to get you off to a happy day. The site you are considering (42° N Lat) has a stand of beautiful, tall, old pines to the southeast, the tops of which are 20 degrees above the horizon from your window. At what time will the sun first shine in your window on December 21?

  Chart: 42° N Lat.

Month XII.

Sun path XII crosses altitude circle 20 degrees at 10 A.M.

Sorry, the pines have to go.

The heating effect of the sun's rays is expressed as the number of BTUs received per square foot of surface per hour of time, or BTU/sq.ft./hour. To understand the significance of such a number we must gain a feel for the BTU.

A BTU is defined as the amount of heat required to raise the temperature of 1 pound (1 pint, the world around) of water by 1 F°. An Englishman (or anyone else, for that matter) could make a pint of tea with 140 BTUs of heat:

$$\frac{\begin{array}{r} 212°F \ temp \ of \ boiling \ water \\ -72°F \ room \ temp \ water \end{array}}{140° \ F}$$

So 140 BTUs is no small amount of heat.

Table 1 lists the amount of heat in BTU/sq.ft./hour falling on the various surfaces of a building at different times of different days.

Example:
Suppose we live in a house at 46° N Lat which is oriented such that one wall faces due south. Of all the surfaces of the house, which receives the most heat on June 22?
Answer: If the roof were flat and horizontal, it would receive heat at the highest rate — 265 BTU/sq.ft./hour — at noon. The runners-up are the east wall (201 at 8 A.M.) and then the west wall (201 at 4 P.M.). The south wall receives only about one-half the intensity of the east and west walls.

The eastern sun at 8 A.M. evokes pleasant memories of feeling like a cat stretching in the sun, but the western sun at 4 P.M. brings to mind sweat, scorched earth, and drinking too much

Table 1. *Solar Position and Hourly Clear-Day Radiation on Surfaces*

26° N. Latitude

A.M.		Alt.	Bear.	S	SE	E	NE	N	SW	Hor.
						BTU/sq.ft./hour				

June 22

A.M.		Alt.	Bear.	S	SE	E	NE	N	SW	Hor.
6 A.M.	6 P.M.	10.05	111.30		49	113	111	44		22
7	5	22.82	105.97		93	185	168	53		81
8	4	35.93	101.15		113	199	168	39		147
9	3	49.24	96.45		111	176	139	20		206
10	2	62.69	88.83	3	94	131	90			253
11	1	76.15	82.61	9	55	69	43			282
12		87.45	0.00	13	9				9	293

March 21, Sept. 24

A.M.		Alt.	Bear.	S	SE	E	NE	N	SW	Hor.
6 A.M.	6 P.M.	0.00	90.00							
7	5	13.45	83.30	17	117	147	92			36
8	4	26.70	75.80	49	172	194	102			101
9	3	39.46	66.33	80	185	182	72			163
10	2	51.11	52.79	104	171	137	23			213
11	1	60.25	31.44	120	137	73			33	246
12		64.00	0.00	125	88				88	257

December 22

A.M.		Alt.	Bear.	S	SE	E	NE	N	SW	Hor.
7 A.M.	5 P.M.	2.23	62.48	14	29	27	9			1
8	4	13.76	54.88	87	149	123	26			37
9	3	24.12	45.30	138	196	139	1			88
10	2	32.66	33.01	171	199	111			42	131
11	1	38.46	17.65	190	177	61			92	159
12		45.55	0.00	197	139				139	168

30° N. Latitude

June 22

A.M.		Alt.	Bear.	S	SE	E	NE	N	SW	Hor.
6 A.M.	6 P.M.	11.48	110.59		55	124	121	47		27
7	5	23.87	104.30		100	189	168	48		86
8	4	36.60	98.24		121	200	162	29		150
9	3	49.53	91.79		121	177	129	6		207
10	2	62.50	83.46	15	103	131	82			252
11	1	75.11	67.48	29	69	69	29			281
12		83.45	0.00	33	24				24	291

March 21, Sept. 24

A.M.		Alt.	Bear.	S	SE	E	NE	N	SW	Hor.
6 A.M.	6 P.M.	0.00	90.00							
7	5	12.95	82.37	19	115	143	88			33
8	4	25.66	73.90	55	174	191	96			95
9	3	37.76	63.44	90	190	179	63			155
10	2	48.59	49.11	117	179	136	13			203
11	1	56.77	28.19	135	147	72			44	234
12		60.00	0.00	142	100				100	245

December 22

A.M.		Alt.	Bear.	S	SE	E	NE	N	SW	Hor.
7 A.M.	5 P.M.	0.38	62.40	2	5	5	2			
8	4	11.44	54.15	78	131	108	21			27
9	3	21.27	44.12	135	189	131			3	73
10	2	29.28	31.73	173	197	107			47	114
11	1	34.64	16.77	195	179	59			96	140
12		36.55	0.00	202	143				143	150
P.M.		α	β	S	SW	W	NW	N	SE	Hor.

Table 1 *continued*

34° N. Latitude

A.M.		Alt.	Bear.	S	SE	E	NE	N	SW	Hor.
						BTU/sq.ft./hour				
June 22										
5 A.M.	7 P.M.	1.47	117.57		6	17	19	9		1
6	6	12.86	109.78		61	135	130	49		33
7	5	24.80	102.54		106	192	166	43		91
8	4	37.07	95.28		129	200	155	19		152
9	3	49.49	87.10	9	131	177	119			207
10	2	61.79	76.00	32	115	130	69			250
11	1	73.17	55.11	48	83	69	15			278
12		79.45	0.00	53	38				38	287
March 21, Sept. 24										
6 A.M.	6 P.M.	0.00	90.00							
7	5	12.39	81.48	21	113	139	83			31
8	4	24.49	72.11	60	175	187	90			89
9	3	35.89	60.79	99	195	177	55			146
10	2	45.89	45.92	129	186	134	3			192
11	1	53.21	25.60	149	156	71			55	221
12		56.00	0.00	156	110				110	231
December 22										
8 A.M.	4 P.M.	9.08	53.57	66	110	90	17			18
9	3	18.38	43.12	129	177	121			6	59
10	2	25.86	30.65	171	193	101			49	96
11	1	30.81	16.05	196	178	56			99	121
12		32.55	0.00	204	144				144	130

38° N. Latitude

A.M.		Alt.	Bear.	S	SE	E	NE	N	SW	Hor.
June 22										
5 A.M.	7 P.M.	3.32	118.42		13	39	42	20		3
6	6	14.18	108.87		68	146	138	50		39
7	5	25.60	100.70		112	195	164	37		95
8	4	37.33	92.25		136	201	148	8		153
9	3	49.13	82.47	23	141	176	108			206
10	2	60.58	69.06	50	127	130	57			246
11	1	70.61	45.67	67	96	69	1			273
12		75.45	0.00	73	52				52	281
March 21, Sept. 24										
6 A.M.	6 P.M.	0.00	90.00							
7	5	11.77	80.63	22	110	134	79			28
8	4	23.20	70.43	65	175	182	83			83
9	3	33.86	58.38	107	198	173	47			137
10	2	43.03	42.16	140	192	131			6	179
11	1	49.57	23.52	162	164	71			65	207
12		52.00	0.00	169	120				120	217
December 22										
8 A.M.	4 P.M.	6.69	53.12	51	84	68	12			10
9	3	15.44	42.30	120	162	109			8	45
10	2	22.40	29.75	166	185	95			50	79
11	1	26.96	15.45	193	174	53			99	102
12		28.55	0.00	202	143				143	110
P.M.		α	β	S	SW	W	NW	N	SE	Hor.

Table 1 *continued*

42° N. Latitude

									BTU/sq.ft./hour	
A.M.		Alt.	Bear.	S	SE	E	NE	N	SW	Hor.
June 22										
5 A.M.	7 P.M.	5.15	117.16		21	61	65	31		6
6	6	15.44	107.87		74	155	145	50		45
7	5	26.28	98.78		118	197	161	30		98
8	4	37.38	89.19	3	144	201	140			153
9	3	48.45	77.96	37	151	176	98			203
10	2	58.95	62.79	67	138	129	44			242
11	1	67.64	38.62	85	109	68			12	266
12		71.45	0.00	92	65				65	274
March 21, Sept. 24										
6 A.M.	6 P.M.	0.00	90.00							
7	5	11.09	79.84	23	107	128	74			25
8	4	21.81	68.88	68	174	177	77			76
9	3	31.70	57.81	113	200	169	40			126
10	2	40.06	40.79	150	197	129			15	166
11	1	45.88	21.82	173	171	69			73	192
12		48.00	0.00	181	128				128	201
December 22										
8 A.M.	4 P.M.	4.28	52.82	35	57	46	8			4
9	3	12.46	41.63	105	141	94			8	31
10	2	18.91	29.01	157	173	87			50	62
11	1	23.09	14.96	187	167	50			97	82
12		24.55	0.00	197	139				139	90

46° N. Latitude

P.M.		Alt.	Bear.	S	SE	E	NE	N	SW	Hor.
June 22										
5 A.M.	7 P.M.	6.97	116.78		28	79	84	40		11
6	6	16.63	106.77		80	162	149	49		50
7	5	26.82	96.80		124	199	157	24		101
8	4	37.22	86.15	14	151	201	132			153
9	3	47.47	73.66	51	160	175	87			199
10	2	56.95	57.25	83	149	128	32			235
11	1	64.40	33.33	103	121	68			25	258
12		67.45	0.00	110	78				78	265
March 21, Sept. 24										
6 A.M.	6 P.M.	0.00	90.00							
7	5	10.36	79.09	23	103	122	70			23
8	4	20.32	67.45	71	172	172	71			69
9	3	29.42	54.27	119	200	165	33			114
10	2	36.98	38.75	157	200	126			22	152
11	1	42.14	20.43	182	176	68			81	175
12		44.00	0.00	190	134				134	184
December 22										
8 A.M.	4 P.M.	1.86	52.65	15	25	20	3			1
9	3	9.46	41.12	87	115	76			8	19
10	2	15.41	28.41	143	156	77			46	45
11	1	19.23	14.56	176	156	46			92	63
12		20.55	0.00	187	132				132	70
P.M.		α	β	S	SW	W	NW	N	SE	Hor.

lemonade or beer — why? The engineers have a measure for this phenomenon: the Sol-Air Index. It refers to the fact that our discomfort or comfort is affected by both air temperature and radiation received. This was shown in the Human Comfort Zone chart. Coincidentally and unfortunately, the maximum daily temperature and the maximum radiation on a vertical surface occur together at 4 P.M. on a summer day. At 8 A.M. the air is still cool and the dew is just leaving the grass.

Example:
At 42° N latitude, which surface receives the greatest radiation on December 22? Now the situation is entirely reversed! The south wall is the highest (197 BTU at noon), the horizontal next, and the east and west receive only a piddling amount.

What are we to conclude from Table 1? *You* may face the road even if it means facing north, but I won't. For another reason, I also won't face west. I will build my house with a lot of windows facing south and southeast, very few to the west (just enough to see the sunset), and perhaps one small window on the north (just to see what's happening). In addition, I will use the sun's noon altitude to determine the roof overhang, which will 1) shade 100 percent of the south window in the summer, and 2) shade none of the south window in the winter.

In the summer, the sun will greet and warm me from the east, be out of sight during midday, and beat fruitlessly on my nearly blank, heavily insulated west wall. In the winter, the sun will still greet me in the east and provide maximum heat through the morning and early afternoon. In the late afternoon the house will already be warm to the point of requiring ventilation and we can afford to accept less radiation.

The winter wind can blast from the north and northwest, but will find no windows to infiltrate.

In the spring and fall we require moderate amounts of sun and that is precisely what we get; as the sun climbs higher and higher, the overhang blocks an increasing percentage of sun.

WIND

Wind is second only to the sun in its ability to modify human comfort. We call unwanted or unpleasant air motion "wind"; pleasant air motion is a "breeze." Nothing can make a site more unpleasant than an eternally blasting winter wind. On the other hand, nothing is more welcome on a hot summer day than a cooling breeze. Is it possible to have one without the other? We will find, to a large degree, that it is. But first we need to find the unmodified (macroclimatic) direction of both winter wind and summer breeze.

The weather bureau and a lot of airports maintain records of wind velocity and direction. If you find that too bothersome, I suggest two alternatives: 1) Daily newspapers usually carry predicted wind velocity and direction in their forecasts (most libraries keep back issues and an hour in the old papers will tell you what you need) or 2) ask some of the old-timers in the neighborhood. It's very rare that a person living in an old, uninsulated house is unaware of the direction of, at least, the coldest wind.

Illustration 25 shows a *wind rose* actually recorded for a site in coastal Maine. The graphs show: 1) direction from which the wind blew (origin of the arrows that terminate at the center); 2) velocity of the wind (thickness of the arrow and); 3) percentage of times blowing from a certain direction (length of the arrow seg-

JAN, FEB, & MAR

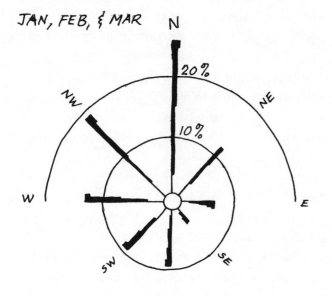

APR, MAY, & JUN

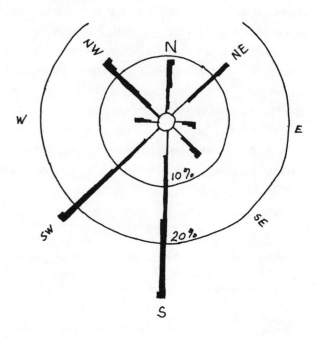

JUL, AUG, & SEP

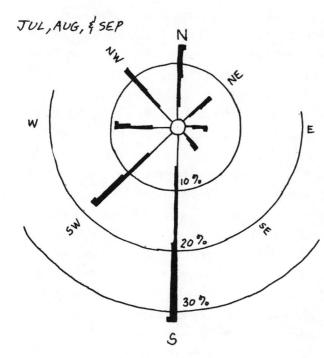

OCT, NOV, & DEC

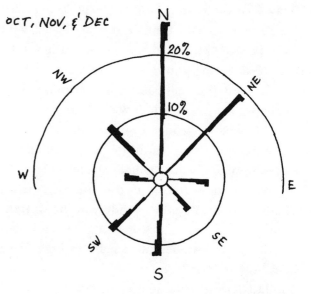

WIND SPEED AND DIRECTION

1 – 4	
5 – 9	
10 – 19	
20 – 29	
30 – 39	

SPEED MPH

25

ment). It is clear from the charts that the cold winter wind blew from the north and northwest and that the cooling summer breezes came from the south and southwest. This wind rose is very typical of the eastern United States. Since the two directions lie about 120° apart, we stand an excellent chance of blocking one and admitting the other. There are two means open to us. The first is (or should be) used by planners and consists of laying out housing units so that the open streets act as channels for ventilating breezes, while in the other direction the closely spaced houses block the force of the winter wind. The second can be used by the individual and consists of capitalizing on the natural effects of trees and shrubs.

Microclimate

SITE SLOPE

We have seen that the relative amounts of solar radiation received on the various surfaces of the house are a function of the altitude of the sun. The sun puts out a constant amount of radiation throughout the year. Thus, the variation in the amount received on a given surface is mostly a function of the angle between that surface and the direction of the sun's rays. The more nearly the rays fall perpendicularly on the surface, the greater the amount of radiation received. The heating effect of the sun on the ground follows the same rule. That is why the macroclimatic spring arrives earlier in lower latitudes. But the vegetation on the ground does not distinguish between altitude of the sun and slope of the

ground. Both angles determine the angle at which radiation is received. The sun's altitude increases roughly 1 degree per 70 miles displacement to the south. Therefore, ground sloping 5 degrees to the south receives the same radiation as level ground 350 miles to the south.

At the same latitude, the arrival of spring (as defined by vegetation) will be observed six weeks earlier on a 10-degree southern slope than on a 10-degree northern slope. That is why the Greeks derived the word meaning climate from the word for slope.

"Cold air falls" is the same as saying "hot air rises." Put your hand at the bottom of a window on a cold winter's night and you will feel the cold air falling just as if it were a waterfall. This phenomenon is acted out on a much larger scale in hilly country. The surface of the ground loses heat by radiating to the clear black night sky. The air in contact with the ground gives up its heat by conduction to the ground. Thus we initially have a uniform layer of cold air along the ground. Cold air, being heavier, seeks a lower level and begins to flow from the adjacent areas into the valley. This results in a "pond" of cold air filling the valley.

The end result is: top of hill — small air motion, ground cold; midslope — maximum air motion generating turbulence that mixes the air layers, resulting in a relatively warmer ground surface; valley — stable cold air pond with coldest ground surface. That is, counter to our intuition, that sheltered little valley will be the coldest spot in the neighborhood on a clear night. The earliest and latest killing frosts always arrive first in the valley, next on the hilltop, and last of all on the south-sloping hillside.

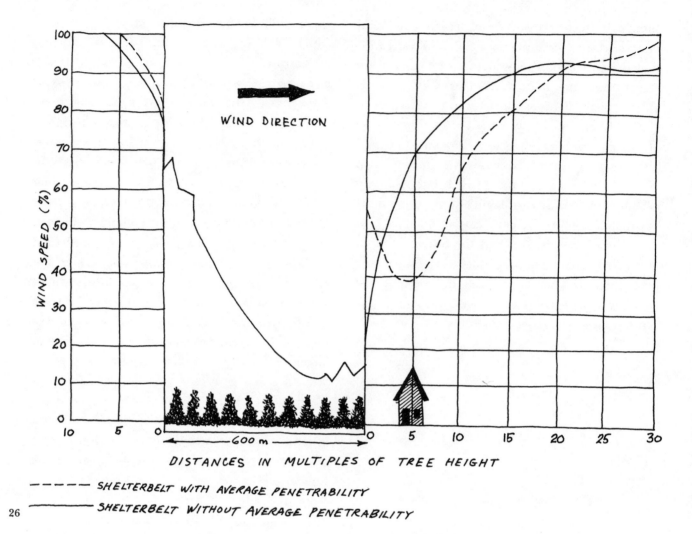

WIND DIRECTION

WIND SPEED (%)

100
90
80
70
60
50
40
30
20
10
0

10   5   0   ←—— 600 m ——→   0   5   10   15   20   25   30

DISTANCES IN MULTIPLES OF TREE HEIGHT

– – – – – SHELTERBELT WITH AVERAGE PENETRABILITY

———— SHELTERBELT WITHOUT AVERAGE PENETRABILITY

26

FOREST EFFECTS

The microclimatic effects of trees is the subject of a little-known document, *Plants, People, and Environmental Quality,* the discovery of which had the same electrifying effect upon me as my first encounter with Rex Roberts's *Your Engineered House.* The magnitude and diversity of weather modification achieved within and by the forest extends far beyond simple wind attenuation. In fact, the forest attenuates without exception every single climatic variable.

Wind in the Forest: Illustration 26 shows a shelter belt of trees and wind velocity both upwind and downwind of the belt. The graph shows a drop in wind velocity within and immediately downward of the belt that is proportional to the density of the trees. Note, however, that a medium-density has a longer downwind effect than a high-density planting. With an open lower level the wind attenuation is small both upwind and within the shelter belt, but great downwind. All of the many charts in the publication show a small upwind attenuation (negligible at an upwind distance of five tree heights) but a strong attenuation downwind (maximum at a downwind distance of about five tree heights with medium density). The principle is clear. If the winter wind and the summer breeze blow from nearly opposite directions we should place our house as below.

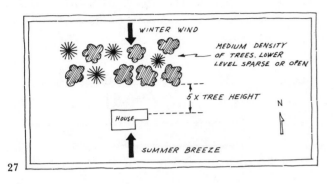
27

If the winter wind and summer breeze lie at about a right angle, then the situation below will be more effective.

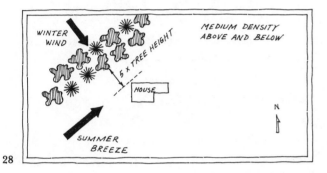
28

The forest can extend your comfort zone by reducing extreme summer temperatures. At the lower end, the primary winter effect is the reduction of wind. Wind chill tables indicate that a 30° F, 15 mph wind outside the woods is equivalent to 9° F. However, at the same time, the wind would be reduced to 4 mph in the woods, giving a chill factor of 28° F — a difference of about 20° F!

Tests on two identical, typical houses in Nebraska were performed to find the effect of a shelter belt on fuel bills. The house in the woods required 30 percent less fuel. Calculations show that an identical house of average leaky construction requires twice as much heat when

exposed to a 20 mph wind as with a 5 mph wind.

Light in the Forest: There are three major sources of visible radiation that may enter your house. The strongest source is the direct radiation of the sun. A surprising amount, however, is received as secondary diffuse radiation from the sky. The third source is reflected glare from the ground surface. On hazy days diffuse radiation may actually exceed direct radiation. At the beach or on the ski slopes, reflected radiation almost doubles total radiation. In the summer, trees may be used to block direct radiation completely. The attenuation of diffuse radiation depends upon the percentage of the sky covered. The reflected glare from trees and the forest floor beneath is almost nil.

In the winter we want all of the radiation we can receive. While evergreens have the same blocking effect year around, deciduous trees will pass most of the winter direct and diffuse light and, with snow on the ground, also allow a considerable amount of reflected glare.

We therefore wish to have a few deciduous trees close to the house. In Illustration 29 the trees to the south block the high noon summer sun. The tree to the west blocks the western 4 P.M. hot summer sun, and the shelter belt provides a welcome early sunset. If we wish, we can have a tree to the east to block late morning

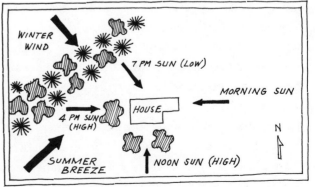
29

sun also. It is possible to find deciduous trees which will thus block 90 percent of the summer sun, but only 10 percent of the winter sun. We can also provide a roof overhang, but only trees are capable of blocking diffuse radiation and also providing a dark, cool, glare-free front field of view.

Air in the Forest: Plants absorb pollutants by their respiration in addition to the carbon dioxide they utilize in producing oxygen. The hairy surfaces of leaves trap an incredible amount of dirt, which is returned to the ground by rain. *Plants, People, and Environmental Quality* quotes a study performed in a German city showing a dust concentration of 3,000 dust particles per air volume in streets planted with trees, but 10,000 to 12,000 particles per volume on treeless streets. Another study showed 2-1/2 acres of beech forest removing 4 tons of dust from the atmosphere each year. The natural freshness of the forest air is something impossible to duplicate, the claims of bathroom aerosol air-fresheners notwithstanding!

Temperature and Humidity: Plants add water vapor to the air by a process known as evapo-transpiration. A large amount of heat is absorbed in evaporating water. Part of the dramatic cooling observed immediately after a thunderstorm on a hot day is due to the evaporation of water from the hot surface of the ground. A tree is a gigantic natural air conditioner. It is estimated that a mature maple tree is capable of evaporating 10,000 gallons of water in a year! Because of the shading of radiation, blocking of wind, and evaporation of water into the atmosphere, the microclimate of the forest is one of controlled humidity and temperature. During the day the forest humidity is higher and temperature lower than in surrounding open

areas. On a clear night, due to radiational cooling, temperatures in open spaces drop sharply with accompanying dew or frost, while in the forest the tree canopy blocks the loss of radiation. Thus, the forests provide relatively stable microclimates.

In Chapter 15 we will discuss the mechanism of frost penetration and we will see that the depth of frost depends upon the insulative value of the ground cover. As a rule of thumb, the maximum depth of frost penetration in the forest is one-half that of surrounding open areas, providing the forest soil is undisturbed. The thick layer of organic debris, known as humus, acts as an insulating blanket.

FINDING SOUTH

You may have concluded by now that the points of the compass are important to you and your house. Here are three ways to find the true direction of south. If you only wish to be sure of catching maximum solar radiation, your orientation to south is not extremely critical. Most experts on solar-tempered houses feel that any orientation between due south and S 30° E is sufficient in the northern half of the United States.

By Magnetic Compass: A magnetic compass needle points to magnetic north, which differs from true north at any locality by an amount known as the magnetic deviation. Illustration 30 shows compass deviations across the United States as of 1965. The deviation changes slowly with time and the exact figure for your town can be easily gotten by calling a local land surveyor. For example, the deviation for Portland, Maine, is found to be 17° W. Go to the site with a good compass. The compass needle will point 17 degrees counterclockwise from true north.

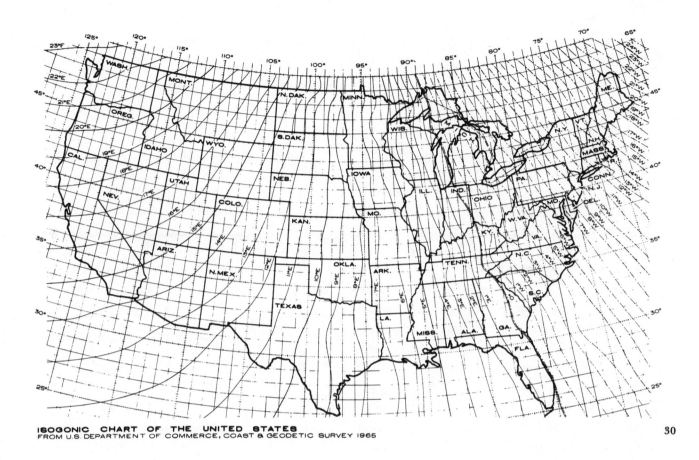

If I rotate the compass until the needle points to N17W, then the compass itself points true north. It is my experience that true north can be found with a cheap compass to about ±5 degrees, good enough for solar orientation. However, if you're beginning to take this orientation business to heart, you'll be interested in the following methods.

By the Noon Shadow: If you have time and patience, but no money, south can be gotten to within about 1 degree by marking the shadow of a vertical object at solar noon. Solar noon is, by definition, the time at which the sun is due south. The object casting the shadow may be an object suspended on a string (plumb bob) or a tall stick held vertically using a carpenter's level. The only remaining problem is finding the standard time that corresponds to solar noon for your site.

Procedure for finding solar noon:

(1) From Illustration 30 find your longitude in degrees and your time zone. The central longitudes for the time zones are: Eastern 75° W, Central 90° W, Mountain 105° W, and Pacific 120° W.

(2) If on standard time, start from 12 o'clock. If on daylight saving time, start from 1 P.M.

(3) Subtract 4 minutes from step 2 for every degree of longitude east of the central longitude, or add 4 minutes for every degree difference if west of the central longitude.

(4) Subtract equation of time correction from Illustration 31 for the appropriate date.

Example:
Portland, Maine, June 21

(1) Time zone — Eastern;
     central longitude   = 75° W,
     longitude of site 70-1/4° W.

(2) Noon Eastern Daylight
     Saving Time   = 1 P.M.

(3)	Add (75 – 70-1/4) X 4 minutes	–	0:19
(4)	Equation of time June 21	=	+0:01

Local time for solar noon	=	12:42 P.M.

By Transit: If you're a real nut like me, you attach cosmic significance to the orientation of your house. My house faces south, so that precisely at solar noon the sun grazes the east and west walls. The pitch of my roof overhang is the same as the noon altitude of the sun on December 21, so that the sun's rays graze the rafters. By marking the shadow of the overhang on the floor, I have a calendar. If you wish to develop the sundial aspect to its limit, you will want to use a transit in laying out your foundation. You'll also want to read *Sundials, Their Theory and Construction,* by Albert E. Waugh. In Waugh's book you'll find a table giving the equation of time for every day of the year, allowing the calculation of local solar noon to within 15 seconds. At the appointed time, one "shoots" the sun with the transit (obtainable at most large tool rental stores). When the transit is lowered to the horizon, anything on the cross hairs is exactly true south of the transit.

(CW)

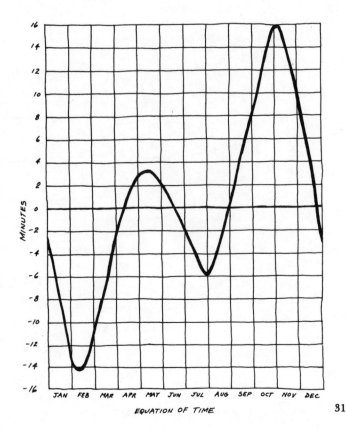

EQUATION OF TIME

31

6

What

Either you own some land, or you don't.

If you do, you must focus your planning on how that bit of real estate can best be utilized so its natural characteristics enhance, protect, and support the shelter you want to build.

If you must search for and select your land — and you have a choice — then the selection should be judged by which of the parcels available can best enhance, support, and protect your shelter.

Landowner or not, you cannot make these judgments effectively unless you have your head in shape. You must already have decided the fundamentals: the reasons why you want to move (which become guidelines for the kind of place you want to move to) and the sort of place you want to live.

I had a fisherman partner and close friend during my years as a commercial fisherman. Hurricanes were routine autumnal visitors in our part of the world, and whenever the warnings came through in those Septembers, Jim and I would have to move our boat, skiff, and dory

to the places we considered safe in any storm. The first of my fishing Septembers, I was surprised when Jim took the 42-foot "big" boat up a small saltwater creek, around a sharp turn in a marshy canal, and into a small cove — a place so confining that we could reach the large oak trees on the shore with our bow and stern lines. Once we had finished the job of making the boat fast, it was obvious that a hurricane would have to obliterate the entire county before our prize possession could be significantly damaged. Yet, looking at the cove casually, I had always assumed it would be too shallow to float a 42-foot boat. Jim had made no assumptions; he had measured the depth.

"Pick your spot, Cap," he said. "Always pick your spot."

The criteria for that particular spot were: enough water to float the boat; protection from winds around the compass; places to tie the four lines from each of the craft's quarters; and an absence of other watercraft that could break loose and ram ours.

I never forgot that episode; when Jean and I

chose our piece of land, it was a spot picked because it best conformed to the criteria we had established. It is in the community where we want most to live, and in the part of that community we find most appealing. It is hidden from the road; it offers a fine view to the south and pine woods on the north.

We were able to get the place, however, because its owners had decided it was too far from the social center of town. Their criteria were different from ours. The difference is that they hadn't quite gotten their heads in shape before they made their purchase. We are delighted that they hadn't, but — wherever possible — you would do well to avoid having to buy one piece of land before you learn what qualities you do or don't want.

The "back" land on our place is damp. The water table is high and the marine clay subsoil does not allow for absorption. Some folks might even call that part of the place a swamp — and they do. In our view, the swampiness is an asset rather than a liability. It means that intense development of the area is quite unlikely, and the individuals in our particular shelter are in agreement that solitude is good and housing developments aren't always.

Pick your spot, Cap. Measure your land with your heads and feet, not a yardstick. Think about it, walk around it, get to know its natural facts. Your shelter is not likely to work well unless you and your land are friends, good ones. And you can't get to be friends unless you get reasonably intimate and honest with each other.

It depends on who you are. If you don't want to build a road through the woods or across the fields to get to your home, then you'll have to buy a parcel that has a good house site hard by the public road. If you don't want to worry about drilling or digging a well to supply your shelter with water, then you'll have to locate near "city" water — whatever that may be.

The land is the sea your shelter will sail on. There is no need for you to take that long voyage on uncharted waters. You can learn the essential identity of your land by making yourself aware of its natural characteristics; you'll have to do it before you can site your shelter so it's in harmony with nature; how much better to do it before you make the purchase!

Prehistoric man — a "natural" being if ever there was one — understood the four keys to the land's essential identity. Earth, air, fire, and water — those are the fundamentals every Neanderthal kept in mind when selecting a cave. Is the cave floor (earth) rocky and wet, or sandy and dry? Is there an opening large enough for good ventilation (air)? Is fuel available for a fire? How far must we travel for water?

The same four keys can still unlock your land's secrets. Is the earth clay, sand, or rock? Will it grow grass, trees, or a garden? Are the trees already there an asset? Will the earth support your home's foundation?

How is the air? What are the prevailing winds, the humidity, the extremes of heat and cold? How much noise will the air bring your way? Can you utilize the wind as an energy supplement? What protection — if any — does the land offer you from the winds of winter?

The cave dweller could frame the same questions today, but he might have some problems with fire. Earth and air may not have changed substantially in the last five thousand years, but fire has. From a small and local blaze on a rocky hearth that gave out heat for warmth, light to chip stone axes by, and flame enough to sear a mammoth steak, fire has been transported and transformed. Now it burns hundreds of miles away not to cook or char, but to generate electricity — the stuff that can do so many jobs that fire never did. It helps communicate, illuminate, cook, sew, and wash; it freezes, cools, and bakes. Most of us find it difficult to get along without. Remember how fire has changed when you consider your land. Ask yourself where your electricity will come from, how your communications can (or cannot) be maintained. Decide whether you want to generate your own "fire" and, if so, how, and how much?

Fire is heat, fire is the telephone, fire is the sun and solar energy, fire is electrical energy, fire is your light and fuel bill. Consider your land and your proposed shelter with these kinds of "fire" in mind and you and the land will get to know each other much better.

The same goes for water if you remind yourself that whatever water comes in to a shelter must also go out. If you are not near a "city" water line, then you'll not only have to develop a water source of your own, but also a water disposal system. For that, you'll need to know more about earth and fire.

Water is also rain, snow, frost, fog, condensation, humidity, and the stuff that your shelter may be bothered by if you build in a natural drainage route, or fail to do a decent caulking job around your windows. Water and air combine in storms to bring horizontal rain, which it is well to be protected from. Water makes puddles, water fills cellars, water can either carry your waste away or it can flood your septic tank.

Think about water when you walk your land; think about air, earth, and fire. Take one of them in your mind each day for four days when you visit your land and ask yourself all the questions each brings to mind. At the end of the four days you will be a great deal closer to comprehending the essential identity of your land.

(Walking your land can be adventurous as well as informative. I met a moose on one of the first explorations of our site. Later, a friend who had come to share another tour stepped into what seemed to be a bottomless pit and sank up to his thigh in soft mud. A bit of digging and probing turned up a natural spring. We later built the kitchen over it. In addition to being assured of a natural emergency water supply, we discovered that the overflow pipe from the spring could also be used as a drainage artery. That discovery has saved us considerable anxiety during cloudbursts and other such meteorological excesses that might otherwise have flooded the front yard, etc.)

Once your site information seems clarified and reasonably accomplished, go back to the fundamentals and rethink the character of the shelter you want and why you want it. Take the land you are considering, work it into the shelter recipe you have created, and give the combination some cerebral kneading for a week or so. What you will be doing, in effect, is evaluating the land that you now know and understand in terms of the place you may want to build there, a place you should also have begun to know and understand.

If you've done the proper amount of thinking, walking, and pondering, if you haven't tried to hustle yourself into going against your better judgment, if you are truly picking your spot, and the combination of shelter and site still seems to work, then (if you do not own the land) you should buy it. If you own the land, but have not been committed to building on it, then commit.

To help you in the struggle that must precede commitment or rejection, here are some basic informational insights about earth, air, fire, and water.

(JNC)

7

What —
Some Details

In looking for land you will probably come across what the realtor or developer calls an "improved site." Whether it has been truly improved by the bulldozer and chain saw is dubious. But the term "improved" refers to the provision of those services that most of us now consider absolutely necessary to existence: road, water, electricity, telephone, and sewage disposal. These items are secondary in importance to the quality and utility of site micro-climatology, but they may consume a quarter of your budget. In comparing land prices, compare *real,* or ultimate, prices; that is, the sum of land, road, water, electricity, telephone, and sewage costs. Land at $1,000 an acre may, in the end, cost less than land at $200 per acre.

Your Road

The important considerations in the cost of a road are length, subsoil, and drainage pattern. A subsoil of sand, gravel, or bedrock, if well drained, requires no foundation. Clay or poorly drained land requires heavy gravel stone foundations. The cost per foot of road consists, essentially, of the cost of hauling those materials. Drainage pattern is a consideration that too often escapes the eye of the novice. Picturesque little brooks (the ones always mentioned in the real estate ads) become raging torrents about April of each year. A road crossing the little brook without provision for handling the maximum annual flow may be washed out every year. *Permanent Logging Roads* (see Bibliography) has tables for estimating required culvert sizes based upon area of land drained and maximum daily precipitation.

A stretch of man-made road is actually a long dam, and water normally flowing downhill, but now intercepted, will run alongside the road to a low point. At the low point water will accumulate until it crosses the road, washing out the surface. At every low point a culvert is required. Standing water is the downfall of the dirt road, for passing vehicles mix the dirt and water deeper and deeper. A road should be carefully graded to shed water immediately: the long-term success of a road absolutely depends upon this ability to shed water.

The shortest path between two points is a straight line; but don't plan to exceed a grade (rise/run) of 10 percent. Experience has shown that grades over 10 percent are dangerous in winter, impossible to plow uphill, and suffer by surface erosion.

Much privacy can be gained with a little curvature in the approach to your house, yet many contractors build driveways in a perfectly straight line on the assumption that, since the shortest path between two points is a straight line, any deviation would cost a great deal more. Let's examine that. Illustration 33 shows three drives. Driveway a is 400 feet long and, at $2.50 per foot, costs $1,000. Driveway b has a total length of only 418 feet and so costs a mere $45 more. Driveway c has a markedly greater length and cost, but actually produces no more privacy than b. Furthermore, the power com-

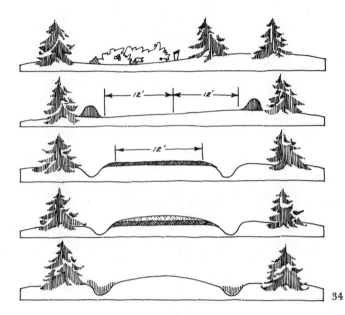

34

pany will find it impossible to follow the sharp curves of drive c and will balk at the increased cost of line and poles. You will find yourself looking straight at the road through a 40-foot-wide hole in the woods resembling a transmission line.

Most communities have several earth-moving contractors (look under "Excavation Contractors" in the Yellow Pages). Get estimates and opinions on the above problems. The reliable contractors will all be close in price and have the same opinion, one based upon solid experience. After you have chosen your contractor, avoid the temptation to pressure him into using less material. You'll pay in the long run. A road with too little foundation is little better than none at all.

The best time to build a road is late fall, before spring construction. Most contractors won't build a road during the spring thaw, and the period immediately after is extremely busy. Without a good road you may find moving building materials consists mainly of pulling vehicles out of the mud. Avoid heavy vehicle traffic (well drillers, concrete trucks, electric

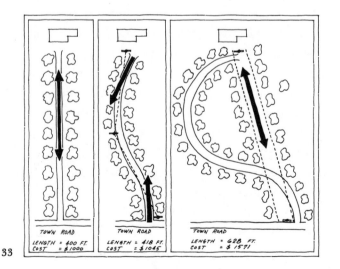

TOWN ROAD
LENGTH = 400 FT.
COST = $1000

TOWN ROAD
LENGTH = 418 FT.
COST = $1045

TOWN ROAD
LENGTH = 628 FT.
COST = $1571

33

company trucks) over your new road until the dry season if at all possible.

Illustration 34 shows how a simple dirt road should be built.

Your Water

Water occurs in the ground in three ways, as shown in Illustration 35: as *surface water* running or ponded at the surface, as *ground water* that has percolated downward through the soil to reach the saturation level (water table), and as *aquifer water,* independent of the water table, flowing in permeable or fractured bedrock.

Surface water, of course, is easy to tap, but sources uncontaminated by humans or industry are pretty rare these days. This is a very ques-tionable source of drinking water without chemical treatment. (And chemical treatment is pretty questionable itself.)

Ground water may or may not be reached with a dug well. This water is more likely to be uncontaminated due to the septic action of aerobic bacteria as the water percolates through the soil. Due to the labor involved in digging a well, you would be wise to consider a dug groundwater well only under the following conditions: depth limited to 15 to 18 feet (reach of backhoe); good groundwater indicators (year-round spring or stream nearby); no known source of pollution. Have the spring water tested by your state department of health.

Illustration 36 shows a method of building a dug well.

The drilled well is best left to the professional. It is the most expensive but also the most reliable and the least likely to be contaminated. Such a well can only go where the drilling truck can go, and state or local plumbing codes specify

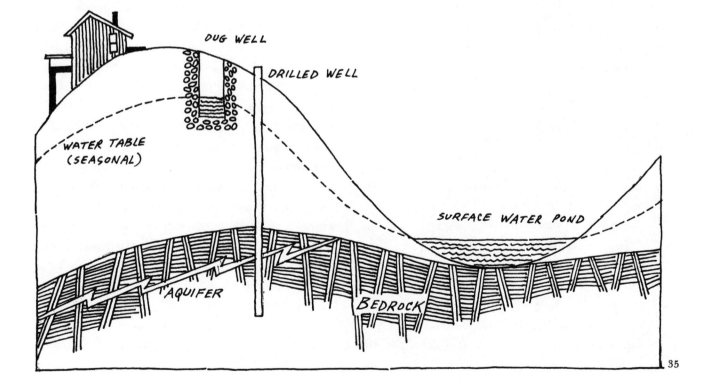

35

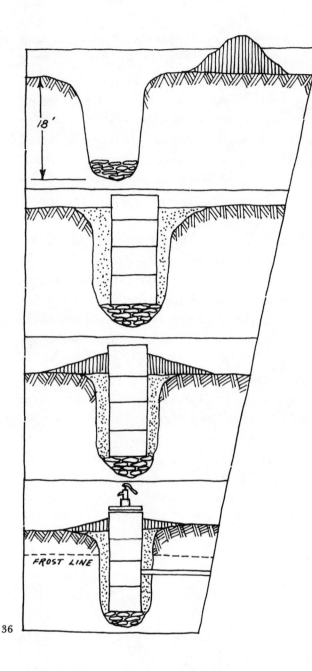

36

linear foot drilled plus price per foot for casing pipe used in unconsolidated overburden; or 2) fixed price to find water — the amount of which decreases with depth. If your site is close to existing wells (not over 1/4 mile) you can obtain the same odds as the driller, who has drill records, by polling your neighbor's depth. The question is, can your budget afford a gamble of $2,000 to $4,000?

Everyone has heard of dry drilled wells. The effect on a nervous stomach of watching a drilling rig sinking at the rate of $5 to $7 per foot can be well imagined. An acquaintance decided he couldn't take it. The day the drilling rig appeared he packed his family away for a vacation. A week later he returned to find the rig still parked in his yard. His heart pounding about as fast and hard as the drill rig, he shouted above the din, "How far?" "Five hundred and sixty and not a drop" came the detached reply. Seeing that this visibly disturbed the home-owner (appearance of tears), the driller pulled up his drill. As a gesture of goodwill the rig was moved 50 feet and started down again, this time at no charge. Water was found at 50 feet. This story illustrates no character flaw in the well driller, but the uncertainties inherent in the whole business.

How much water is enough? Most people imagine one gallon per minute as the merest trickle. You couldn't wash your Cadillac with it. Plumbing codes usually specify a waste water handling capacity of about 200 gallons per person per day. Actual figures show an average use of more nearly 100 gallons per person per day. Households with hand pumps and privies manage nicely on 10 gallons per day total. Anyway,

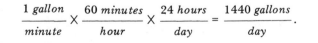

$$\frac{1\ gallon}{minute} \times \frac{60\ minutes}{hour} \times \frac{24\ hours}{day} = \frac{1440\ gallons}{day}.$$

a minimum separation of well and septic systems — usually 100 feet.

The level at which aquifer water is found is *somewhat* consistent over a local area. However, since the water is flowing in cracks that cannot be seen, an unsettling percentage of holes miss the large cracks. Most well drillers, with several hundred thousand dollars invested in equipment, can afford to gamble with you and offer contracts on two bases: 1) price per

Twenty-six 55-gallon drums per day is enough water for fourteen average people. A 6-inch drilled well with a 100-foot drawdown (difference between static or equilibrium water level and bottom of pipe) has 150 gallons in storage — enough water for seven baths. Three families can live on a one-gallon-per-minute well and never run dry.

Getting Power from the Wind

The wind is a potential source of energy that derives from the thermal driving forces of solar energy. If your sole interest in electricity is lowest dollar cost and you are reasonably close to an existing power line, read no further. If, however, environmental issues come far above budget in your priority list, then read on.

The available energy in wind varies with the third power of the wind speed:

$$Power = Constant \times Velocity^3.$$

That is, compared to a 15 mph wind, a 10 mph wind has only $(10/15)^3$ or 30 percent as much power and a 7-1/2 mph wind has only 12 percent as much. You may have the impression that the wind blows *at least* 15 mph all the time at your site, but we humans are prone to exaggeration when it comes to describing wind. Unless you live on a mountaintop or an ocean island, I can almost guarantee that your yearly average wind speed is much less than 15 mph. The average for the entire United States is about 7-1/2 mph. Locations with a 10 mph average include Key West, San Francisco, Indianapolis, Pittsburgh, and Dallas. Atlantic City, Block

Island (Rhode Island), and Tatoosh Island (Washington) get 15 mph.

Fifteen mph is significant because that is the minimum average wind speed for the presently most competitive commercial 115 VAC windmill system to compete with Reddy Kilowatt at 8 cents per kwh. Due to the velocity cubed law, the cost of electricity from the same mill at other speeds would be: 10 mph, 27 cents per kwh; 7-1/2 mph, 64 cents per kwh!

Why are windmills so expensive? If Volkswagens were handmade at the rate of several dozen per year, the humble "bug" would probably retail at $30,000. No doubt the windmill will become increasingly competitive as fuel prices go up and mass production costs go down, but at this moment the relative cost of wind-generated electricity is prohibitive for the average site. Purchase of windmills by environmentally conscious people who can afford the sacrifice will hasten the day when we can all afford energy-self-sufficient homes that do nothing to pollute the earth.

Many homemade windmills have been described in popular magazines over the past few years. If your spare time would otherwise be spent building stock cars, then go ahead. Perhaps you will advance the state of the art a bit. Otherwise please consider two points: (1) A windmill must operate 24 hours a day for years on end. Will the components of the homemade mill stand up without expensive maintenance? (2) Many of these mills have low outputs at average windspeeds. Compute the power produced at *your* average windspeed and compare it to the amount of power you presently consume. If you are average your household consumes around 500 kwh per month.

If you do live on the coast or on a high bald hill, or if the nearest utility line is five miles from your site, and you are willing to severely restrict your consumption, then read *Electric*

Power from the Wind, a little booklet available from Solar Wind Publications, P.O. Box 7, East Holden, Maine 04429.

Reddy Kilowatt

Local power companies have historically had policies promoting increased use of electricity. First they advertise "safe, clean, 100 percent efficient electric heat" and then plead that public demand requires additional generating facilities.

In spite of signs of impending change, most power companies have policies that encourage the development of new electric lines. An example would be (check with your local utility line department):

> 5-year contract —
> 2,000-foot new line maximum per year-round home
> 1,000-foot new line maximum per seasonal home
> minimum monthly charge = 1-1/2¢/foot/month.

In other words two families at the end of a 3,500-foot line would be charged a *minimum* of

3,500 × .015 = $52.50 *per month total, or,*
$26.25 *per month each for five years.*

Communication

Electric and telephone companies share the cost of poles. Generally the electric company places the poles and the telephone company pays a use fee per pole. The telephone policy is not quite as liberal. After one or two free poles per family, you pay the one-time use fee of about $70 per pole.

Sewage

Waste disposal is such a delightful subject that an entire chapter is devoted to it. Just in case you are already negotiating over a site, read Chapter 23 and visit your local plumbing inspector before signing a thing. Nothing has the potential of driving up site costs more than inadequate soil conditions for sewage disposal.

(CW)

The Materials

8

With Your Own Two Hands

You have determined your climate, shaped your comfort zone, discovered and deciphered your land, and designed your shelter to fit your psyche and your site.

Now comes yet another internal confrontation: the struggle to convince yourself that you can do much of the construction work with your own two hands.

Arrogance and ignorance can be great allies. They are enlisted, however, at high cost.

For most of the five thousand years humankind has been around, shelters have been constructed by the humans who intended to dwell within them. Kings, potentates, emperors, and other such royalty may have employed architects and arranged for slaves to design and construct palaces, etc., but millions of men and women through the ages have put their own roofs over their heads.

Not until the Industrial Age was well on its way as a dominant presence in the United States did the "masses" become convinced of the notion that contractors, builders, plumbers, architects, inspectors, electricians, carpenters,

masons, and linoleum cutters were essential to shelter construction. And thus the "housing industry" was born, grew, and grew some more, until it became one of the giants of American industry — essential, so it is claimed, to the welfare of the nation's economy, and obsessed with a compulsion for exponential growth.

Matters of national import are involved (whether you realize it or not) in your grapple with the concept of do-it-yourself home construction. Conditioning to which you have been subjected since childhood is at work. You have been told many times that electricians alone can understand a light bulb, that carpenters alone can cope with hanging a door, or that a contractor alone can comprehend the details and manage the people it takes to build a house. It is a kind of heresy for you to even consider flouting a national truism.

It is easier if you are arrogant. "If farmers, their families, their friends and their oxen could put up the graceful barns that still rise like century-old cathedrals along every country road, then surely, equipped with power tools, a college education, a superlative mind, and a determination to succeed, then surely I too can build my own home." That, or a similar litany, was what I told myself every day for the year our shelter was in its planning and design stage.

By the time the morning arrived when the first shovel had to be pressed into the ground to start the foundation, I had reached a compromise that I considered a victory. My family and I did design our new home; my sons and daughters and their high school contemporaries did do much of the construction and carpentry work; we did not use a professional contractor, but instead we enlisted the help of a friend who was "handy" in many ways, but had never before built a house of any kind.

The heating system, the plumbing work, and the major part of the electrical wiring was done by professionals. I told myself I had to employ their services because our time frame for building was so restricted; I simply didn't have the free hours (I told myself) to learn how to become a plumber, electrician, and heating engineer. That was partially true; I was also anxious enough about failure — and aware enough of its possibilities — to delight in the discovery that the time shortage also meant I would have to use professional help.

My arrogance could sustain me only through the design phase and the basic carpentry, overseeing, and improvising. It was my ignorance that got me even that far.

I kept looking at those barns — sailing like great gray ships on their hand-hewn granite foundations — and telling myself that a home (especially ours) was little more than a barn with partitions and spaces for people instead of machinery and livestock. Posts were the timbers that were vertical, beams were horizontal, and rafters were on a slant. I knew that much. I could tell by looking at the barns. When I

looked a bit more carefully, from underneath, I found even larger beams supporting the floor.

"There's nothing to it," I told myself and my family. "We put the beams here, the posts here, the rafters here, the windows here, the doors here . . ." I kept going in that arrogant, ignorant, and naïve way until I had the entire house sketched on paper. I invested in a copy of *House Beautiful* so I could copy its system for drawing plans. In my mind's eye, I put beams across concrete foundation posts, I elevated frames onto that base (just as the frames for the old barns had been elevated), and I roofed it, plumbed it, wired it, and heated it, and there it was — on paper and in a three-dimensional model built from posterboard — our shelter.

It is, after all, so simple, I told myself. A house is nothing more than a collection of nails, thousands of them, driven — one after the other — each one, into a collection of wooden members, each one measured and cut, one after the other. Whatever openings are left must be filled with either doors or windows, and there you have it: a shelter.

You will have to believe that our home was built on such basic premises. It truly was. The reason I became convinced we could do it was because I never let the concept get complicated. I reduced the project, constantly and consistently, to the lowest common denominator for that particular day. My ignorance, of course, was overwhelming, my arrogance even more so.

With those givens, the shelter that resulted is a triumph of good fortune and excellent advice generously given. I never failed to listen to those generous voices. A visiting and curious architect, looking at the arrangement of windows on the south side of our home, suggested that the structure might suffer from racking if we did not install some sort of cross-bracing. Until he uttered the word, I had remained ignorant of racking and its potential. We did, however, install the cross braces; now visitors want to know where we got the clever design.

Another visitor who knows much of forces and moments (see the following chapter) suggested that our roof beams might not successfully support six feet of wet snow. The beams were already in place; yet I did not want to be sitting in the living room on the day six feet of wet snow cascaded onto the rug. We invented a cable truss on the spot. The trusses are a prominent feature of the place; visitors always ask what they are; the best description of their presence is "audacious." The word was bestowed by one of those many visitors.

What I am saying, in brief, is that we did not know what we were doing (in large measure) when we set about designing and building our own place. But we produced a shelter that has served us well, aesthetically and functionally. The primary reason for success is our belief that we could do the job. We convinced ourselves, therefore we did.

In addition to such matters as after-the-fact braces and one-of-a-kind cable trusses, there are other penalties for arrogance and ignorance. Of these, anxiety is the most persistent and penetrating.

It has been three years now since we moved into the place. During those years we have been visited by record rains, hurricane winds, snow, sleet, frozen drains, chipmunks, raccoons, red squirrels, and below-zero winter nights of excruciating length and intensity.

During every one of these natural phenomena I was contorted by doubt, galvanized by anxiety. Whenever the wind blew whitecaps on the bay, I wondered if it would flip the house over and roll it through the woods, like so much sagebrush across a desert. When the rain fell for more than an hour, I would wait, trembling, for the house to lift from its foundation and float off like Noah's ark without its passengers.

In the cold, I would expect the glass to crack; in the heat, I would watch for a melting roof.

The house haunted my dreams. In the dark hours my arrogance could not sustain the fears my ignorance had spawned. Even though the house was there, protecting me and my beloved family, I could not rid myself of the fears that somehow, somewhere, along that endless procession of plans, nails, and fitted boards, I had made an error of such statuesque proportions that it was only a matter of time before the entire structure collapsed around me like the buildings Clark Gable escaped from during the earthquake scene in *San Francisco.*

Such are the long nights of an arrogant and ignorant home builder. And yet the place is here. It has been built by my sons and daughters and their friends. We did design it. We did much of the work with our own two hands. We sought to reach a harmony with natural forces, to be warmed by the sun and cooled by the winds. We wanted to be involved, to be free of the tyranny of the specialists, to put our roof over our own heads and to live happily under that roof, sustained by the self-confidence that would accrue from having done just that.

We did that. We did it because we believed we could. Primed by arrogance, sustained by what we did not know — rather than what little we did — and made painfully anxious because of that combination, we nevertheless built our shelter. And the project succeeded, both by our own measure and the measure of those we care for and respect.

Imagine, then, if we could do that with so little knowledge and such sketchy information, what you can do with the physics and metaphysics that are gathered here for you. Not only can you become fully capable of doing the work with your own two hands, but you can do it with knowledge rather than guesswork, facts rather than the fancies of optimism. You can go great steps further and unravel the details of wiring, the strategy of plumbing, and the basics of home heating.

You can do it so much better than we. The information you need is all here, gathered and prepared for you. The process can be so different, so informed, so anxiety-free. However, it must also be something of the same. Like us, you must arrive at the fundamental faith in yourself that will sustain you and say to yourself: "I can do it with my own two hands."

Of course you can. Your ancestors have done it throughout history, thousands of years before there was such an entity as the "housing industry."

Prepared with information on such matters as "forces and moments" you won't need my arrogance or ignorance; nor will you suffer my concomitant anxieties.

(JNC)

9

Forces and Moments

Most people assume that the design of a house frame is either based upon past knowledge of workable solutions or a difficult mathematical problem to be dealt with only by those with engineering degrees. Neither is true. For those of us over thirty, only first-year high school math is required. For the younger generation, grammar school math will suffice.

Nothing will please you more than to *truly design* your house frame. No other aspect is as liberating to design as the ability to analyze and understand the stresses on a house. And nothing will save you more money.

Throughout this book we will use certain unit abbreviations regularly. A number is meaningless without the accompanying unit. For example, the quantity 10 is completely meaningless unless we are dealing only with pure numbers. Ten apples, 10 square miles, 10 cubic feet, 10 pounds, mean completely different things.

Always include the unit names in any of your calculations. When the unit name of the answer to a calculation makes no sense, it tells you that the answer makes no sense; it tells you that your analysis must be reexamined.

Abbreviation	Unit
in.	inch
ft.	foot
yd.	yard
sq.in.	square inch
sq.ft.	square foot
cu.in.	cubic inch
cu.ft.	cubic foot
cu.yd.	cubic yard
lb.	pound
in.lb.	inch-pound
ft.lb.	foot-pound
lb./cu.ft.	pounds per cubic foot
psi	pounds per square inch
psf	pounds per square foot

Forces

Each of us has an intuitive definition of force: it is that which causes things to move. However, we realize that sometimes a given force is not sufficient to produce motion. Therefore the definition of force must be more precise.

Force is the phenomenon that, if unbalanced, causes a change in motion (either starts motion, increases motion, decreases motion, or changes the direction of motion). No matter how far out our conception of a house may be, it should not move! Therefore we are interested only in *balanced forces.* That simplifies the problem immensely.

One example of a balanced force is you standing on the ground. Your weight is a force downward on the ground. What we most often ignore is the fact that the ground is pushing back with an equal (balanced) force. To make this more intuitive: Get a friend to push his hands against your hands without moving. Now have him hold your feet while you stand on your hands. Is the ground pushing up?

You do not move because the upward force of the ground balances the downward force, your weight. If suddenly the ground disappeared there would remain only the downward force and down you would go in response to the unbalanced force.

In buildings forces acting downward are called *loads.* The equal and opposite forces acting upward in reaction to the loads are called *reactions.* We will always express forces in pounds (lb.).

Moments

Illustration 39 demonstrates a second sort of motion. We know that the reaction upward at the seesaw pivot is equal to the sum of the two

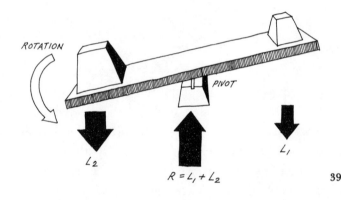

39

downward loads because the seesaw as a whole does not move up or down. But the seesaw can move — by rotating. That is because the loads and reactions do not act at the same place. To deal with this problem we use the concept of the moment: *A moment is a force times its distance from a pivot. The distance is taken at right angles to the force, and the moment is positive if it acts to cause clockwise motion, negative if counterclockwise.*

In Illustration 40 the two moments are equal and of opposite sign. Moment 1 acts to produce clockwise rotation, while Moment 2 acts to

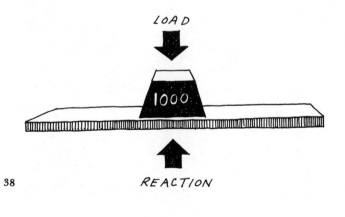

38

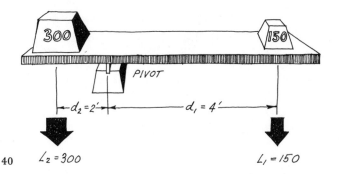

40

$Moment\ 1 = L_1 \times d_1 =\ 150\ lb. \times 3\ ft. =\ 450\ ft.lb.$

$Moment\ 2 = L_2 \times d_2 = -300\ lb. \times 2\ ft. = -600\ ft.lb.$

$$Sum = -150\ ft.lb.$$

The sum of the moments is negative. Therefore, by the sign convention, the seesaw will rotate counterclockwise.

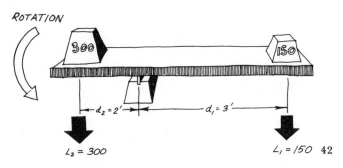

42

produce counterclockwise rotation. The moments are balanced and no rotation occurs.

$Moment\ 1 = L_1 \times d_1 =\ 150\ lb. \times 4\ ft. =\ 600\ ft.lb.$

$Moment\ 2 = L_2 \times d_2 = -300\ lb. \times 2\ ft. = -600\ ft.lb.$

$$Sum =\ \ \ \ 0\ ft.lb.$$

Illustration 41 shows that the moments may be balanced and produce no rotation even when the seesaw is initially tipped.

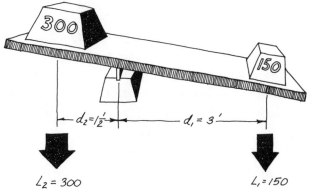

41

All motion of rigid bodies consists of either translation (motion in a straight line) or rotation (motion around a pivot), or a combination of both. Therefore, all we need to know about the possible motion of a house is contained in the following two rules:

Rules of Moving Houses

(1)
If the sum of the reactions equals the sum of the loads, then no translation occurs.

(2)
If the sum of the clockwise moments equals the sum of the counterclockwise moments, then no rotation occurs.

$Moment\ 1 = L_1 \times d_1 =\ 150\ lb. \times 3\ ft.\ \ =\ 450\ ft.lb.$

$Moment\ 2 = L_1 \times d_2 = -300\ lb. \times 1\frac{1}{2}\ ft. = -450\ ft.lb.$

$$Sum =\ \ \ \ 0\ ft.lb.$$

Illustration 42 shows how rotation may be predicted when moments are unbalanced.

At this point you probably think I'm crazy. What's all this about translating and rotating houses? You may never have seen a house that has translated or rotated, but believe me, they do. Mine did. The reason you never see them in this state is that the owners are so embarrassed that the houses are immediately restored and the subject is never mentioned again. The feeling and reaction are akin to having your pants fall down in public.

Some years ago, immediately after receiving a Doctor of Science degree from the citadel of technology, MIT, I purchased a rundown shore-front cottage in Maine. The original foundation was a cedar post affair with a full perimeter skirt of boards. Time, insects, and a seasonal high water table that rose several inches above ground level had taken their toll on the post bottoms and the cottage listed a noticeable amount. Physical labor in the sea air proved invigorating and I spent a full month building a stone foundation under the cottage. Time ran out after three sides had been enclosed, leaving the highest side open. Three sides, however, represented considerable stability (a damn sight better than before, anyway!), and I left quite pleased with myself.

The next April we received a cryptic note in the mail from the proprietor of the local general store. "Maybe you should come look at your place." Period. The next weekend we drove up, fantasizing all sorts of terrible scenes such as roof shingles blown off, water stains on the ceiling, broken windows, etc.

The scene that greeted us was one of those that evokes the feeling "I'm glad that's not my house!" The house had *translated* 10 feet and *rotated* 30 degrees. As if that weren't bad enough, the bottom third of the chimney was in place, the top third had fallen through the porch roof, and the midsection had apparently rattled around inside the building.

In retrospect, my analytical mind theorized that the wind sweeping into the open foundation found no exit and exerted an upward pressure, thereby reducing the effective weight of the building. The horizontal force of the wind on the side of the building was then stronger than the resisting friction and the house slid off the foundation. My confrontation with that cottage and my first reading of Rex Roberts may claim equal credit for my interest in houses.

Loads

DEFINITIONS

Dead load — *The weight of the building itself.*
Live load — *The additional weight due to occupancy of the building.*
Snow load — *The vertical weight of snow on the roof.*
Wind load — *Forces directed at right angles to building surfaces by the pressure of the wind.*

DEAD LOAD

The dead load is calculated as the weight per area in psf of each building surface (roof, floors, walls, foundation). The total load is then found by multiplying each by the respective surface area. The calculation may be done exactly by using published values of building material density. However, the softwood conventionally used in houses has a dry density of very nearly 25 lb./cu.ft. That is, for each square foot of building surface, the dead load is about 2 pounds for each inch thickness of wood used. If you bother to calculate, you will find that every wall and floor has a dead load of very nearly 5 psf, and roofs with asphalt roofing (shingles, selvage, roll roofing) weigh an average of 7-1/2 psf.

LIVE LOAD

The National Building Code (NBC) dictates the live loads that must be assumed for residential construction:

first floor (interpret as social areas)	40 psf
second floor (interpret as bedrooms)	30 psf
attic (interpret as unoccupied)	20 psf

You may object that 40 psf implies a crowd of one person for every 3 square feet or eighty people in a 12′ × 20′ living room; but if twenty couples start dancing to Leon Russell you'll have the equivalent — and it only takes once!

Also remember that an unfinished attic planned to become a bedroom, just in case, is a bedroom by the code.

SNOW LOAD

From Illustration 43 you can find the snow load in psf exceeded statistically one year in ten for your area. This is derived from the measured average snow pack on the ground. I know that most older homes, being poorly insulated, never accumulate half of this load, but of course your house will be so well insulated that the snow will remain on your roof as long as on the ground. People who build with steep roofs in order to shed snow are cheating themselves. Eleven inches of dry snow have the same insulating effect as 3-1/2 inches of fiberglass.

Let's say you live in the north and the map says your snow load is 40 psf. This corresponds to a depth of 5 feet of powdery snow or 1 to 2 feet of wet snow — an event that is not likely to escape your attention. However, let's also

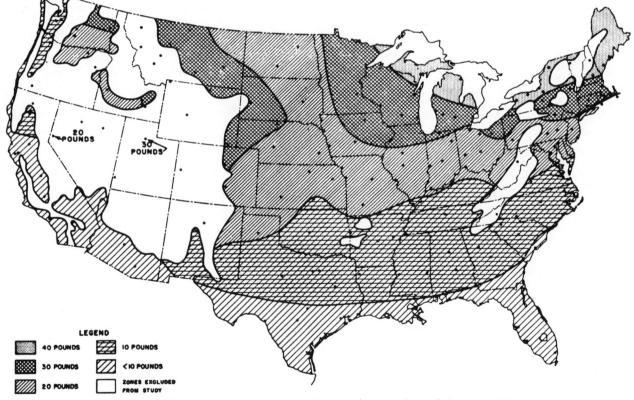

LEGEND

40 POUNDS   10 POUNDS

30 POUNDS   <10 POUNDS

20 POUNDS   ZONES EXCLUDED FROM STUDY

Estimated weight in pounds per square foot of seasonal snowpack equaled or exceeded one year in ten   43

assume that your friend is getting married in Toledo on the very date of the big storm. Again, it only takes once, so let's go with the recommended load.

Steeply pitched roofs do not retain snow as well as flat roofs and so a reduction in snow load is allowed for pitch. Check your local building code for this feature.

The rule cited by the NBC allows a reduction of any snow load over 20 psf by the amount $(S/40 - 1/2)$ psf for each degree of pitch over 20 degrees.

Example:
For a snow load of 40 psf and pitch of 45° we are allowed to subtract:
$$(S/40 - 1/2) \times (45° - 20°) \, psf = 1/2 \times 25 \, psf$$
$$= 12\text{-}1/2 \, psf$$

So the design snow load is 27-1/2 psf. Before you get too excited, however, there are lots of other reasons for keeping the pitch down around 20 degrees.

WIND LOAD

There are three ways in which a house can move in high winds: *vertically,* by pressure under the floor and airfoil suction over the roof; *horizontally,* by horizontal pressure against the effectively lighter structure; *overturning,* by a combination of horizontal and vertical moments. An exact analysis of the forces involved would be extremely complicated, so simplified methods have evolved. I know of at least four different methods and have chosen the simplest for you. Once again, if your town has a building inspector, verify his acceptance of this method.

Vertical: A roof is similar to an airfoil. The increased velocity of air flow over the roof surface results in a lower pressure. Therefore,

calculate the total uplift on the building as 12 psf times the horizontal area occupied by the building. In addition, if the house has a large overhang or eave, calculate the uplift against the overhang separately at 30 psf because the wind is pushing from below as well as pulling from above. In case of a double overhang, consider only the upwind one. Recalling my experience, a house with a foundation wide open on one side but blocked on the other three sides is, in effect, one giant overhang!

Horizontal: The horizontal force of the wind is calculated as 20 psf on the exposed vertical surface (the windward wall). The resistance to sliding is the net weight of the building (the dead weight of the building minus the vertical uplift) times a coefficient of friction between building and foundation of 0.6.

Overturning: Remember that the measure of tendency to rotate (overturn) is a sum of moments — forces times distances. Here the forces are the vertical uplift, horizontal wind force, and weight of building as discussed above. The distances are taken from the center of each force to the pivot, or point about which the building would most likely rotate. The center of a force is always taken as the exact midpoint of the house surface involved. For example, horizontal wind force is assumed to act at a point halfway up the wall. The building weight is assumed to be concentrated at the center of the building, etc. The sum of the moments caused by the wind is known as the *overturning moment* and the moment of the building weight is known as the *restoring moment.* The building will overturn, of course, if the overturning moment exceeds the restoring moment. The foundation may be included in the building weight provided the building is anchored to the foundation. In addition, the weight of earth

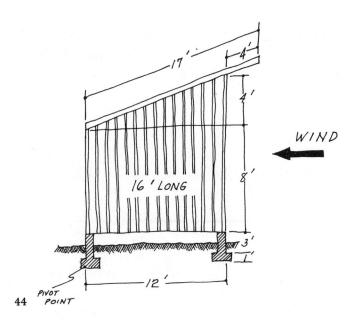

44

directly above the footing may be included when the footing is tied to the foundation. When the foundation is included, the pivot point becomes the intersection of the outside leeward building line with the footing bottom. If there is any question as to the location of the pivot: 1) assume that the wind will blow against the highest side of the building (always assume the worst!); and 2) imagine the wind as a giant hand pushing against the house. The house will act as if it were hinged at the lowest downwind point.

Example:
Unit weights

Walls, floors	5 psf
Roof	7½ psf
Concrete	150 lb./cu.ft.
Earth	120 lb./cu.ft.

Building weight	*pounds*
Floors 2 × 12' × 16'	
× 5 psf	1,920
Walls 560 sq.ft.	
× 5 psf	2,800
Roof 17' × 16'	
× 7.5 psf	2,040

6,760 lb. subtotal

Foundation weight	*pounds*
Posts 6 × 1' × 1'	
× 3' × 150 lb./cu.ft.	2,700
Footing 6 × 2' × 2'	
× 1' × 150 lb./cu.ft.	3,600
Earth 6 × 3' × 3 sq.ft.	
× 120 lb./cu.ft.	6,480

Total   12,780 lb. subtotal

19,540 lb. *Deadweight*

Uplift   12 psf × 12' × 16' = 2,304 lb. above floor
        +30 psf × 4' × 16' = 1,920 lb. overhang

Total   4,224 lb.

Sliding   Horizontal Force = 20 psf × 12' × 16'
                = 3,840 lb.

Friction =
weight – uplift   = (6,760 – 4,224) × 0.6
                = 1,522 lb.

Overturning Moments
M_1 = 2,304 lb. × 6' = 13,824 ft.lb. (lift on roof)

M_2 = 1,920 lb. × 14' = 26,880 ft.lb. (lift on overhang)

M_3 = 3,840 lb. × 10' = 38,400 ft.lb. (wind on side)

Total   79,104 ft.lb.

Restoring Moment = 19,540 lb. × 6' = 117,240 ft.lb.

In the example we first calculate the deadweight of the building alone. This is noted as a subtotal. Second, we calculate the weight of the foundation, including the earth directly over the footings, and record it as a subtotal. Third, we add the weights of building and foundation.

Uplift: The roof directly over the floor area of the house is subject to an uplift of 12 psf. Note that the area is taken as the horizontal projection. This force is assumed to be concentrated at the center of the roof, excluding the overhang. The overhang is subject to a separate, higher pressure of 30 psf. Again the force is assumed to act at the center of the overhang. Since the uplift on the roof is greater than the weight of the roof, we reach *Conclusion #1*: The roof must be securely fastened to the walls, especially at the overhang. The weight of the building without foundation is greater than the total uplift, and so it is not yet established whether foundation anchor bolts are required.

Horizontal Sliding: The horizontal force is computed by multiplying horizontal wind pressure by the area of vertical wall exposed to the wind. The resistance to sliding is the friction of the net weight of the building against the foundation and is computed as building deadweight minus uplift, all times the coefficient of friction 0.6. The resistance to sliding is found to be less than the horizontal force. *Conclusion #2:* The building must be anchored to its foundation.

Overturning: The overturning moment is the sum of three simultaneous moments: 1) lift on roof, 2) lift on overhang, and 3) horizontal force. Moment is defined as force times distance. By the definition of moment, the distance must be taken at right angles to the direction of the force. That is, for a vertical force the distance is taken as the horizontal distance from the pivot to the horizontal center of the force. For a horizontal force the distance is measured vertically from the pivot to the center of the force. We assume that the various forces all act at the midpoints of the surfaces. *Conclusion #3:* The building anchored to the foundation will not overturn, the restoring moment being greater than the sum of the overturning moments.

The validity of the assumption of forces acting at midpoints is proven in two ways: What other assumption is more valid? And imagine yourself balancing a sheet of plywood on one finger — where does it balance?

What kind of wind produces pressures of 20 psf? The highest wind velocity you'll ever encounter other than on the top of a bald mountain or in southern Florida. Are we allowed to reduce the wind pressure when located in the forest? Not according to the NBC, but then the people who wrote the NBC probably don't know how nice it is in the woods.

Bracing Against Racking

Racking is the distortion of a building under horizontal shearing forces. The shearing forces are usually those due to the wind pressure against exterior walls.

Shear may be produced in all of the walls of a building due to the wind pressure on adjacent walls. However, we worry most about the racking of walls that consist largely of windows. Racking in a window wall stresses the integrity of the weather seal and ultimately may cause infiltration around the window. Of course, it may also break the glass in extreme winds, particularly if an imperfection in glass or frame causes a stress concentration. As a rule of thumb, a window wall should have a continuous plywood sheet from top to bottom plates at some point in each wall.

To design this plywood sheet:
(1) Determine the magnitude of shear in the wall. This is computed as 1/4 × wind pressure in psf × adjacent wall area in sq.ft. divided by the attached plate length.

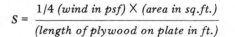

$$S = \frac{1/4\ (wind\ in\ psf) \times (area\ in\ sq.ft.)}{(length\ of\ plywood\ on\ plate\ in\ ft.)}$$

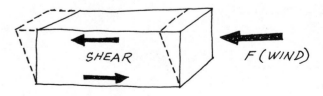

SHEAR   F (WIND)

45

Table 2. *Shear Resistance of CDX Plywood Wall Panels*

Plywood Thickness	Nail Size (common or galvanized box nails)	Nail Spacing Around Perimeter			
		6"	4"	2-1/2"	2"
1/4"	6d	180 lb./ft.	270	400	450
3/8"	8d	260	380	570	640
1/2"	10d	310	460	670	770

The answer is in lb./ft.

(2) Refer to Table 2 for the required nail size and spacing for the computed shear and thickness of plywood used.

Example:
Shear in a window wall

$$S = \frac{1/4 \times 20 \, psf \times 200 \, sq.ft.}{4 \, ft.} = 250 \, lb./ft.$$

From Table 2 we find: 1/4" plywood, nailing @ 6" o.c.

3/8" plywood, nailing @ 4" o.c.

1/2" plywood, nailing @ 6" o.c.

If the wall is higher than 8 feet, either special-order a piece of plywood of full length or make

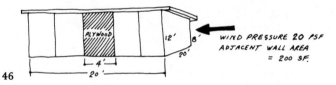

46

your own plywood sheet by overlapping and gluing several pieces. Be sure the plywood is dry, use plastic resin or elastomeric glue, and apply

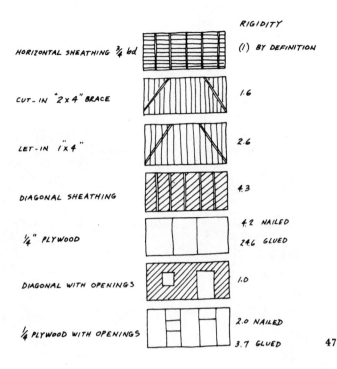

	RIGIDITY
HORIZONTAL SHEATHING ¾ bd	(1) BY DEFINITION
CUT-IN "2 x 4" BRACE	1.6
LET-IN "1 x 4"	2.6
DIAGONAL SHEATHING	4.3
¼" PLYWOOD	4.2 NAILED / 24.6 GLUED
DIAGONAL WITH OPENINGS	1.0
¼ PLYWOOD WITH OPENINGS	2.0 NAILED / 3.7 GLUED

47

pressure by piling lots of other sheets of plywood on top.

Tests have demonstrated the superiority of plywood over any other common building material in providing racking resistance. Illustration 47 shows the relative rigidities found in one such test. Subsequent testing has shown that buildings with either Homesote exterior sheathing or Sheetrock interior walls perform as well as horizontal boarded walls with 1" × 4" let-in braces. However, since both materials are relatively subject to nail hole enlargement after repeated loadings, plywood is still superior. Half-inch CDX plywood applied with waterproof glue and 8d annular ring nails is my choice.

(CW)

10

Wonderful Wood

Even though it has been several years since we moved into the home we designed and built "with our own two hands," many of the same thoughts continue to repeat themselves each time I look around it. One of the most frequent of those thoughts is the recollection that, until my thirtieth year, I knew absolutely nothing about wood. I scarcely knew where it came from, much less how to analyze its strengths and recognize its weaknesses. That's what can happen to a fellow who is raised in the city and educated in classrooms.

In that thirtieth year, however, I began my apprenticeship as a commercial fisherman, and wood was what kept us afloat. With scarcely any capital at our disposal, my partners and I had to make do with whatever boats we could acquire at incidental prices or, quite often, merely in return for hauling the wrecks out of a boatyard.

For seven years I restored, repaired, and rebuilt ancient and abused wooden watercraft, and in the course of that education I became acquainted with most of the different kinds of wood, with their vagaries and frustrations as well as strengths and needs. I painted wood, scraped it, sanded it, drilled it, steamed it, caulked it, nailed it, fastened it, and fashioned it into stems, sterns, planks, masts, booms, hatch covers, tables, door frames, tackle boxes, net carriers, engine beds, and clam rake handles. Everything from mahogany, oak, pine, cedar, spruce, teak, cypress, and ash suffered under my clumsy hands.

By the time I departed that career, older, wiser, but still a zealous fisherman, I had learned a bit about wood. I had learned that oak is stringy, heavy, and strong; that one sands with the grain; that large nails driven near the ends of planks will usually split those planks; that spruce is lighter than ash (and therefore better for oars); that even a bit of rot in a mast will weaken it; that wood and water react in wondrous ways; that even relatively delicate lengths of wood, if properly fashioned, can be incredibly strong and resilient enough to resist the battering of awesome seas.

I also discovered and learned to appreciate the

enduring beauty of wood as is, unpainted and unstained. Which is quite likely one good reason why our entire home and just about everything in it is made of wood, as is, unpainted and unstained.

Wood is one of the great, largely unrecognized natural resources of this nation. There are many places on this globe where wood is a rarity, where the burning of oak logs in a picnic bonfire would be considered as wanton as skipping gold coins into a deep lake. But in America, we take wood for granted. Many of us can live thirty years or more without even understanding its origins and its character.

I am no longer quite that ignorant, and I thank my fishermen friends for the insights they helped me gain. From those beginnings I developed a love for wood that still grows. I find sensuality in the texture of a well-sanded, well-oiled plank. I am captivated by the aesthetics of knots, grain, and hue. I am inspired by wood's organic immortality; it never dies. With every change in temperature, with every ray of sunshine, drop of rain, breath of salt fog, or splash of ocean spray, wood responds. It swells, shrinks, curls, changes color; it reacts as if the soul of the tree still locked within it has never departed, but is there, reaching and responding to the presence of moisture just as the parent tree did every day of its life in the forest.

Wood is a renewable and immortal resource. It is never still. Unlike steel, aluminum, plastic, stone, or brick, wood moves. It is alive, it glows, and, if you care for it, wood will shelter you with maintenance-free strength, wood will warm you, wood will please your eye, and wood will wrap you in its immortal strength when you go to your grave. Nature has provided us with no better structural material. Wood needs only to remain dry or be well ventilated if it becomes wet. It's best to keep it dry and venti-

lated, and to rub it with natural oils now and then to replace those it has lost in the process of separating from the tree.

Wooden doors, frames, and furniture buried with the ancients of Egypt several thousand years ago are still as strong, functional, and as lovely as the day the tombs were closed. So you needn't worry about "old" wood. If you understand grain, can judge which kind of wood might be best for what sort of purpose, if you know the varieties and can sense something of the wonder of wood's immortality, wood will do well by you and your home.

Let me tell you one wood story.

The new owners of our former home wanted our dining room table when we sold that place, so we left it for them. During the six months we worked before we could move to our new home, I forgot the missing piece of furniture. Our first supper in the new place was eaten from a small picnic table borrowed from a friend. It was not large enough for all the family together.

The next day I rummaged in the lumber pile in front of the still-abuilding house. We constructed the place primarily with planks and timbers taken from two large Maine barns which, in turn, had been built more than a century before. Their present owner wanted them down; he was going to bulldoze and burn them. That seemed such a waste to us. We wanted our place to minimize waste rather than stress consumption. The wood worked well.

But some of it had been difficult to find a function for. There were heavy hemlock planks, for example, that had come from the floors of the cow stalls. More than an inch thick and better than 22 inches wide, the planks had been battered by hooves and soaked with urine and cow manure.

To me, in my impatience to get a family table built, the hemlock seemed made to order. I took three planks. Their massive, collective breadth totaled more than 5 feet. With a nail driven in the center, a string, and a bit of chalk, I drew a circle across that breadth, cut along the line with a saber saw, sanded the planks down 1/4 inch, and mounted the three-piece disk atop a square frame built of barn timbers.

The table was meant to be temporary, but it was strong enough so I could dance around its edge, handsome enough so no one wanted to let it go. Rubbed with linseed oil every several weeks, and brushed by scores of hands every day for three years of meals, the hemlock has acquired a lovely soft patina. It glows under candles in the evening, shines in the morning sunrise.

Those hemlock planks served the creatures of the manger for more than a century; now they serve us just as well, and will continue to serve longer than we will live. Wood is always beautiful, always immortal. You only have to know that to know how to treat it. You will never know all the uses it can be put to.

You can, however, make a start by learning a bit more about it.

(JNC)

11

Wonderful Wood — Some Details

Everyone knows that wood has different strengths depending upon whether it is oak or pine; whether it has knots; how big it is in cross section, how long, etc. We basically have three choices in building a house: 1) be supercautious and overspecify (not *overdesign,* for no real design is involved); 2) buy graded lumber from a yard (which is better, but still gives us no credit for proper understanding of the wood); or 3) learn the qualities of wood and optimize our designs around them. Method #3 can save 50 percent of the cost of framing a house.

Forces in Wood

We should be concerned with four sorts of forces in wood, each with a different name according to what each tends to do.

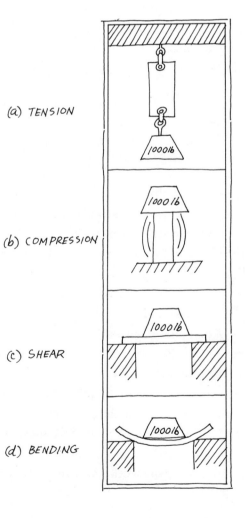

(a) TENSION

(b) COMPRESSION

(c) SHEAR

(d) BENDING

49

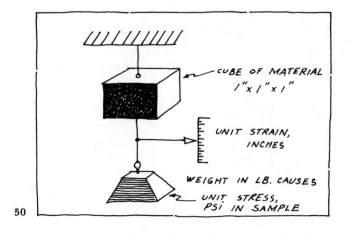

50

STRESS AND STRAIN (Force and Stretch)

Stress in a material is the measure of internal force that resists an external force. The larger the piece of material under stress the more external force it can resist. Therefore, to be able to compare material of varying size, unit stress is defined:

$$f = F/A$$
where f = unit stress in psi
F = total force in lb.
A = area of cross section in sq.in.

According to the atomic theory, most of the volume of even the strongest of materials consists of empty space. Therefore, we may imagine that whenever we apply a *stress,* as small as it may be, there is a resulting *strain.* What follows is true of any material, be it wood, glass, steel, or stone.

If we had an extremely sensitive measuring device we could perform the following experiment (see Illustration 50):

We fashion a cube of material exactly 1 inch on a side. Starting from zero, we slowly increase the stress by adding weight and measure the corresponding strain, or movement, of the pointer. The cross-sectional area, A, of the

cube is 1 sq.in. The results of our experiment are shown in Illustration 51.

Hooke's Law: As we slowly increase the stress from zero, we find that the strain increases proportionately, giving us a straight line. That is:

(Hooke's Law) $$\frac{Change\ in\ stress}{Change\ in\ strain} = E.$$

E is a measure of stiffness or rigidity and has the engineering name *modulus of elasticity.* ("Modulus" is a fancy term for "ratio.") In the straight-line region where Hooke's Law applies, removal of the stress results in the return of strain to zero. That is what "elastic" means. However, in continuing to increase stress we reach a point where not only does the strain increase more rapidly, but removal of the stress results in a permanent strain, or "set." We have at this point exceeded the *elastic limit.* Increasing the stress even further, we ultimately reach a point where strain increases without limit and without requiring further stress (simply put — it breaks). This point is the material's *ultimate strength.*

Inhomogeneities and imperfections in wood lower the strength of wood, as discussed below. Visual grading of lumber for framing purposes essentially consists of specifying an *allowable*

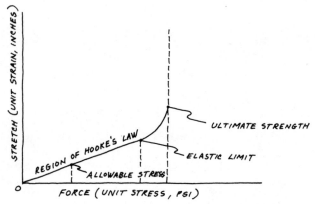

51

stress, or working unit stress, according to visible defects. The ratio of ultimate strength to allowable stress is called the *factor of safety*. The reason why that carport you built last summer withstood the winter's snows in spite of violating the building code is that the factor of safety for wood is usually around eight!

Example:
We have a timber of cross section $8'' \times 10''$ and an allowable unit compressive stress parallel to grain, $C_{/\!/}$, of 1,000 psi. What total load is it allowed to carry as a post?

$$F = C_{/\!/} \times A$$

where F = *total force in lb.*
$C_{/\!/}$ = *allowable compressive stress parallel to grain in psi*
A = *area in sq.in.*
or
$F = C_{/\!/} \times A = 1,000 \ psi \times 80 \ sq.in. = 80,000 \ lb.$

BEAMS IN BENDING

When a piece of wood is subjected to forces that make it bend it is generally known as a beam. Specific examples of beams are floor joists, ceiling joists, and roof rafters. We are interested in two aspects of bending in a beam: Will the beam break? How much will it deflect under a given load?

If I asked you where to place the load in Illustration 52 to most easily break the beam, you would have no trouble in arriving at the middle. And in Illustration 53 you would, I'm sure, say the end. You have already deduced the fact that force and distance are both important in bending. In fact, the quantity we are looking for is an old friend — the moment. Appropriately, its name is now *bending moment*, and it is measured in *inch-pounds*.

Now what are the qualities of a beam that give

it strength? Again you already know, I'm sure. Suppose you have two planks, one of oak and one of pine. In Illustration 54, which one would you rather walk across? The oak one, I suspect.

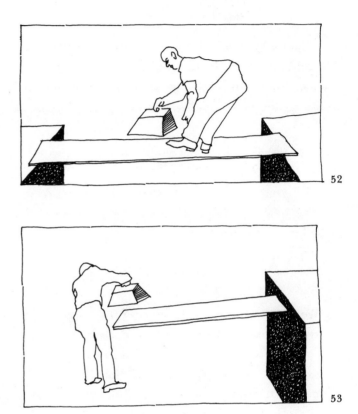

52

53

In other words, there is an inherent difference in strength in bending between wood species. The quantity that expresses this difference is known as f — *the extreme fiber stress in bending*, measured in psi.

There is also a geometric quality to strength. Which way would you lay the plank, on its edge or its face? Rafters and joists are used vertically because they are stronger that way. The quantity that expresses geometric strength is *section modulus, S*. For example, if in Illustration 54 $b = 2''$ and $d = 6''$, the beam would be three times stronger (have a section modulus three times greater) when placed on edge:

$$\text{Section Modulus, } S = \frac{b \times d^2}{6}$$

$$S = \frac{2'' \times 6'' \times 6''}{6} = 12''^3 \qquad S = \frac{6'' \times 2'' \times 2''}{6} = 4''^3$$

(on edge)   *(on face)*

The way in which all of the bending quantities combine is simple. Imagine a tug-of-war between the forces of breaking and the preservers of integrity.

$$\underset{\substack{\text{(bending} \\ \text{moment)}}}{M} \quad = \quad \underset{\substack{\text{(extreme fiber stress} \\ \text{in bending)}}}{f} \quad \times \quad \underset{\substack{\text{(section} \\ \text{modulus)}}}{S}$$

The equality represents the point of breaking. The beam is unsafe when the left-hand side, *M*, is larger numerically.

DEFLECTION

We are concerned with deflection somewhat less than failure in bending since deflection has no relation to safety. You will find "deflection" nowhere in the National Building Code. However, the usual criterion for "bounciness" of a floor is that the ratio of vertical deflection to span of the beam be less than 1/360 under maximum live load. I, personally, use this rule for first floors. *Example:* 1/360 translates into

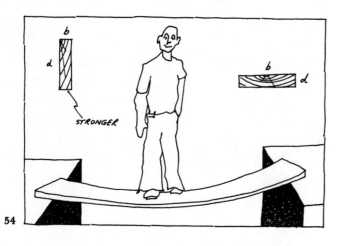

STRONGER

54

1 inch of vertical deflection at the center of a 30-foot beam; 1/2 inch over 15 feet, etc. The quantities determining deflection are the total load on the beam, the span, the *modulus of elasticity, E* (stiffness), and a quantity similar to section modulus but called *moment of inertia, $I = bd^3/12$.*

LOADS ON BEAMS

Loads are usually placed on beams in two simple ways: distributed uniformly or concentrated at one or more points. An example of a uniformly distributed load is the live load of 40 psf on a residential floor. A single post bearing upon a sill is a concentrated load. Before calculating the bending strength of beams in various situations we must learn to find the load carried by each beam. Illustration 55 shows a simple floor consisting of joists spaced at 16 inches on center (o.c.) resting upon a pair of sills, in turn supported by posts 8 feet o.c.

Each joist carries one-half of the uniformly distributed weight of the space on each side. For interior joists this is equivalent to each joist carrying the weight of an entire space. The weight carried by end joists is only half as great, but since they also carry the end walls they are actually doubled in practice. The sills carry the joists, and if there is a large number of uniformly spaced joists, the weight is assumed to be uniformly distributed on the sill.

Example:
Uniformly Distributed Loads
The clear span, *L*, of the joists, is the length unsupported from beneath. In Illustration 55 it is seen that $L = 15'0''$. The clear span of the sills is seen to be $7'6''$.
Joist. The distributed load on a wood floor is the dead load of 5 psf plus the live load (40 psf for first floor). The total load carried by a first-floor joist *over its clear span* is, therefore, 45 psf times the

$$\text{Section Modulus, } S = \frac{b \times d^2}{6}$$

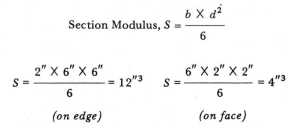

$$S = \frac{2'' \times 6'' \times 6''}{6} = 12''^3 \qquad S = \frac{6'' \times 2'' \times 2''}{6} = 4''^3$$

(on edge)   *(on face)*

The way in which all of the bending quantities combine is simple. Imagine a tug-of-war between the forces of breaking and the preservers of integrity.

M	$=$	f	$\times$	S
(bending moment)		*(extreme fiber stress in bending)*		*(section modulus)*

The equality represents the point of breaking. The beam is unsafe when the left-hand side, *M*, is larger numerically.

DEFLECTION

We are concerned with deflection somewhat less than failure in bending since deflection has no relation to safety. You will find "deflection" nowhere in the National Building Code. However, the usual criterion for "bounciness" of a floor is that the ratio of vertical deflection to span of the beam be less than 1/360 under maximum live load. I, personally, use this rule for first floors. *Example:* 1/360 translates into

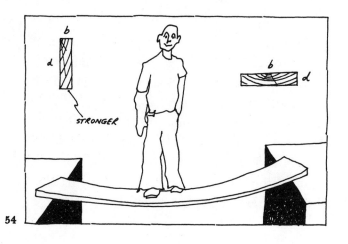

1 inch of vertical deflection at the center of a 30-foot beam; 1/2 inch over 15 feet, etc. The quantities determining deflection are the total load on the beam, the span, the *modulus of elasticity, E* (stiffness), and a quantity similar to section modulus but called *moment of inertia, $I = bd^3/12$.*

LOADS ON BEAMS

Loads are usually placed on beams in two simple ways: distributed uniformly or concentrated at one or more points. An example of a uniformly distributed load is the live load of 40 psf on a residential floor. A single post bearing upon a sill is a concentrated load. Before calculating the bending strength of beams in various situations we must learn to find the load carried by each beam. Illustration 55 shows a simple floor consisting of joists spaced at 16 inches on center (o.c.) resting upon a pair of sills, in turn supported by posts 8 feet o.c.

Each joist carries one-half of the uniformly distributed weight of the space on each side. For interior joists this is equivalent to each joist carrying the weight of an entire space. The weight carried by end joists is only half as great, but since they also carry the end walls they are actually doubled in practice. The sills carry the joists, and if there is a large number of uniformly spaced joists, the weight is assumed to be uniformly distributed on the sill.

Example:
Uniformly Distributed Loads
The clear span, *L*, of the joists, is the length unsupported from beneath. In Illustration 55 it is seen that $L = 15'0''$. The clear span of the sills is seen to be $7'6''$.
Joist. The distributed load on a wood floor is the dead load of 5 psf plus the live load (40 psf for first floor). The total load carried by a first-floor joist *over its clear span* is, therefore, 45 psf times the

stress, or working unit stress, according to visible defects. The ratio of ultimate strength to allowable stress is called the *factor of safety*. The reason why that carport you built last summer withstood the winter's snows in spite of violating the building code is that the factor of safety for wood is usually around eight!

Example:
We have a timber of cross section $8'' \times 10''$ and an allowable unit compressive stress parallel to grain, $C_{/\!/}$, of 1,000 psi. What total load is it allowed to carry as a post?

$$F = C_{/\!/} \times A$$

where F = *total force in lb.*
$C_{/\!/}$ = *allowable compressive stress parallel to grain in psi*
A = *area in sq.in.*
or
$F = C_{/\!/} \times A = 1,000 \, psi \times 80 \, sq.in. = 80,000 \, lb.$

BEAMS IN BENDING

When a piece of wood is subjected to forces that make it bend it is generally known as a beam. Specific examples of beams are floor joists, ceiling joists, and roof rafters. We are interested in two aspects of bending in a beam: Will the beam break? How much will it deflect under a given load?

If I asked you where to place the load in Illustration 52 to most easily break the beam, you would have no trouble in arriving at the middle. And in Illustration 53 you would, I'm sure, say the end. You have already deduced the fact that force and distance are both important in bending. In fact, the quantity we are looking for is an old friend — the moment. Appropriately, its name is now *bending moment,* and it is measured in *inch-pounds.*

Now what are the qualities of a beam that give

it strength? Again you already know, I'm sure. Suppose you have two planks, one of oak and one of pine. In Illustration 54, which one would you rather walk across? The oak one, I suspect.

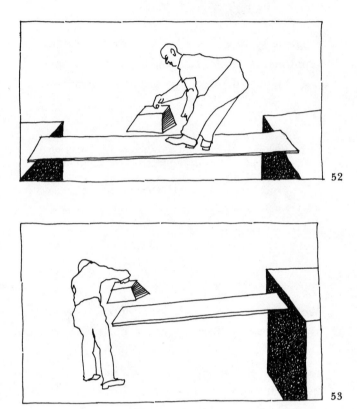

52

53

In other words, there is an inherent difference in strength in bending between wood species. The quantity that expresses this difference is known as f — *the extreme fiber stress in bending,* measured in psi.

There is also a geometric quality to strength. Which way would you lay the plank, on its edge or its face? Rafters and joists are used vertically because they are stronger that way. The quantity that expresses geometric strength is *section modulus, S.* For example, if in Illustration 54 $b = 2''$ and $d = 6''$, the beam would be three times stronger (have a section modulus three times greater) when placed on edge:

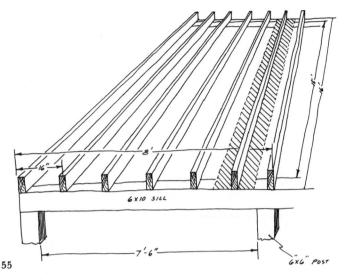

55

area of the rectangle formed by the clear span and the o.c. spacing.

Load W = 45 psf × *15 ft.* × *1-1/3 ft.* = *900 lb.*

Sill. The distributed load on a sill is similarly 45 psf times the area of the rectangle formed by the sill clear span and one-half the joist clear span.

Load W = 45 psf × *7-1/2 ft.* × *7-1/2 ft.* = *2,536 lb.*

Why half of the joist clear span instead of half of the joist length? Because the wall space inside the walls is unoccupied. The weight of the walls is a separate calculation.

Example:
Concentrated Loads
We could also frame the floor using heavy joists 48 inches o.c. Here we have substituted one 6″ × 8″ joist for every three 2″ × 8″ joists. The o.c. spacing is now 48 inches.
Joists. The center joist carries

W = 45 psf × *15 ft.* × *4 ft.* = *2,700 lb.*

Sills. Bending force is applied to each sill now at only one point — the center joist. This is clearly a concentrated load. The load is one-half of the load on the joist.

W = 1/2 × *2,700 lb.* = *1,350 lb.*

Design Cases

The majority of design cases in residential construction are of the sort discussed in the foregoing example. However, true design freedom requires the ability to deal with more challenging possibilities. Table 3 lists six different design cases

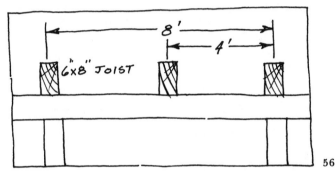

56

with formulae for reactions, deflections, and bending moments. Sometimes a real situation involves two simple cases simultaneously. Nothing could be simpler! If a beam carries two types of loads, the two cases are worked out independently and the reactions, deflections, and bending moments are simply added together.

Table 4 gives values of *f*, fiber stress in bending, and *E*, modulus of elasticity, for nine common wood species. Column I contains the average ultimate strengths as actually measured in the laboratory, using visually perfect or clear specimens. Columns II and III give the corresponding measured shear strengths and moduli of elasticity. Column IV lists the allowable fiber stress in bending for long-term (10-year) loading. Columns V and VI list the corresponding allowable long-term shear strength and moduli of elasticity. Wood has a greater strength over shorter time spans. Columns VII and VIII give the increased values of *f* for load durations

Table 3. *Beam Formulae*

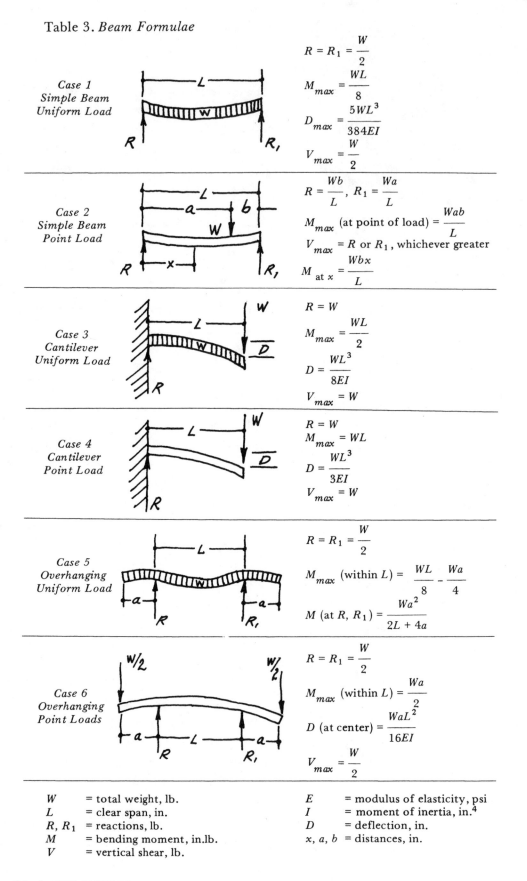

Case 1
Simple Beam
Uniform Load

$$R = R_1 = \frac{W}{2}$$

$$M_{max} = \frac{WL}{8}$$

$$D_{max} = \frac{5WL^3}{384EI}$$

$$V_{max} = \frac{W}{2}$$

Case 2
Simple Beam
Point Load

$$R = \frac{Wb}{L}, \; R_1 = \frac{Wa}{L}$$

$$M_{max} \text{ (at point of load)} = \frac{Wab}{L}$$

$$V_{max} = R \text{ or } R_1, \text{ whichever greater}$$

$$M_{\text{at } x} = \frac{Wbx}{L}$$

Case 3
Cantilever
Uniform Load

$$R = W$$

$$M_{max} = \frac{WL}{2}$$

$$D = \frac{WL^3}{8EI}$$

$$V_{max} = W$$

Case 4
Cantilever
Point Load

$$R = W$$
$$M_{max} = WL$$
$$D = \frac{WL^3}{3EI}$$
$$V_{max} = W$$

Case 5
Overhanging
Uniform Load

$$R = R_1 = \frac{W}{2}$$

$$M_{max} \text{ (within } L) = \frac{WL}{8} - \frac{Wa}{4}$$

$$M \text{ (at } R, R_1) = \frac{Wa^2}{2L + 4a}$$

Case 6
Overhanging
Point Loads

$$R = R_1 = \frac{W}{2}$$

$$M_{max} \text{ (within } L) = \frac{Wa}{2}$$

$$D \text{ (at center)} = \frac{WaL^2}{16EI}$$

$$V_{max} = \frac{W}{2}$$

W = total weight, lb.	E = modulus of elasticity, psi
L = clear span, in.	I = moment of inertia, in.^{4}
R, R_1 = reactions, lb.	D = deflection, in.
M = bending moment, in.lb.	x, a, b = distances, in.
V = vertical shear, lb.	

Table 4. *Average Mechanical Values of Wood Species*

| | Clear Specimen | | | Ave. Allowable, 10 Year | | | 1 Month | 1 Day | Safety Factor |
	Fiber Stress in Bending f(psi)	*Shear Stress S(psi)*	*Modulus of Elasticity E(10⁶ psi)*	*f(psi)*	*S(psi)*	*E(10⁶ psi)*	*f(psi)*	*f(psi)*	*Clear Allowable*
Northern White Pine	8,600	900	1.24	900	80	1.0	1,200	1,300	9.6
Eastern Hemlock	8,900	1,060	1.20	1,200	80	1.2	1,400	1,500	7.4
Eastern Spruce	9,800	1,080	1.34	1,200	100	1.2	1,400	1,500	8.2
Douglas Fir	12,400	1,130	1.95	1,500	120	1.6	1,700	1,900	8.3
White or Red Oak	15,200	2,000	1.78	1,800	190	1.5	2,100	2,200	8.4
Sugar Maple	15,800	2,330	1.83	2,000	190	1.6	2,300	2,500	7.9
Eastern Red Cedar	8,800	—	0.88	1,200	60	0.8	1,400	1,500	7.3
American Elm	11,800	1,510	1.34	1,400	150	1.2	1,600	1,700	8.4
Paper Birch	12,300	1,210	1.59	1,600	190	1.6	1,800	2,000	7.7
	I	II	III	IV	V	VI	VII	VIII	IX

These are average values subject to grading.

of one month and one day. Column IX shows the factor of safety defined as Column I divided by Column IV.

All of the allowable values in Table 4 are for *average* mill-run lumber. Of course in any lot, a few pieces will have much greater than average strength and a few will have a much smaller strength. Later we discuss how to visually grade your lumber by recognizing the effect on strength of various obvious defects. Use of the table assumes an understanding of wood defects and grading.

DESIGNING BEAMS

Each design case is subject to the same procedure. Following the step-by-step procedure below will result in the dimensions of the required beam.

(1) Pick a wood species and grade $f =$ ____ psi; $E =$ ____ psi.

(2) Define case number from Table 3 and write formulae.

(3) Define clear spans, L, and o.c. spacing.

(4) Calculate W, total load, either uniform or concentrated.

(5) Calculate Maximum Bending Moment, M_{max}.

(6) Calculate Section Modulus, $S = M_{max}/f$.

(7) Find breadth (b'') and depth (d'') from table of S, or from formula $S = bd^2/6$.

(8) Calculate Moment of Inertia from b and d and formula $I = bd^3/12$.

(9) Calculate Maximum Deflection, D_{max}. Check ratio of D_{max}/L. Is it greater than 1/360?

(10) If so, recycle, increasing d or b.

We will return below to the example of Illustration 55.

Example:
Joist

(1) Eastern hemlock, $f = 1,200$ psi; $E = 1,200,000$ psi.

(2) Case 1, $M = WL/8$, $D = \dfrac{5\ WL^3}{384\ EI}$.

(3) $L = 15'$, o.c. spacing = 1-1/3$'$.

(4) $W = 45\ psf \times 15' \times 1\text{-}1/3' = 900\ lb.$

(5) $M_{max} = \dfrac{900\ lb. \times 15' \times 12''}{8} = 20,250\ in.lb.$

(6) $S = M_{max}/f = \dfrac{20,250\ in.lb.}{1,200\ lb./sq.in.} = 16.875\ in.^3$

(7) Assuming $b = 2''$, $S = bd^2/6$

$$16.875 = 2d^2/6$$
$$d^2 = 50.625 \ in.^2$$
$$d = 7.11 \ in.$$

Therefore, joist is $2'' \times 8''$.

(8) $\qquad I = bd^3/12 = \dfrac{2 \times 8 \times 8 \times 8}{12} = 85.33$

(9) $\qquad D_{max} = \dfrac{5WL^3}{384EI}$

$$= \dfrac{5 \times 900 \times (15)^3 \times (12)^3}{384 \times 1,200,000 \times 85.33}$$
$$= 0.67 \ in.$$

$$D_{max}/L = \dfrac{0.67}{15' \times 12''} = 1/270$$

(10) If you personally feel 1/270 is too great, re-compute from step 8 using $2'' \times 10''$.

Shear: The six design cases of Table 3 also include formulae for computing vertical shear. Shear is rarely the cause of failure in a beam, but can occur when a beam carries great weight over a short span. An example would be a cantilevered floor joist carrying a wall post, in turn supporting a second floor load plus roof load. *Always check for horizontal shear in short beams or beams with great depth (d).*

Procedure for Checking Shear
(1) Compute Vertical Shear, V, from formula found in Table 3.
(2) Compute Horizontal Shear Unit Stress, H, from

$$H = 3V/2bd,$$

where b and d are breadth and depth of beam cross section in inches.
(3) H must not exceed the Horizontal Shear Unit Stress values for the beam species shown in Table 4.

Span Tables

Case 1 is so often used in floors and roofs that complete span tables for both surfaced (lumberyard) and rough (local mill) lumber are given here (Table 5). The loads of 25, 35, 45, 37-1/2, and 47-1/2 psf correspond to the total loads on attic, second and first floors, and roofs with snow loads of 30 psf and 40 psf.

Important: The span tables are for Case 1 simple beams only, and cannot be substituted for other cases. Cases 2–6 can only be calculated using the Design Procedure.

What Is Wood?

Everyone knows that lumber is essentially sliced tree trunks. Few know about the trees' growth history, the sawing procedure, the seasoning, and how the imperfections influence the strength of the lumber. To optimize the design of our house we have to understand such things.

A tree is a living plant and thus is composed of cells. Wood cells are extremely long and spindle-shaped, running primarily in the direction of the trunk of the tree. When the tree is first cut it is "green," which refers not to its color but to the fact that the cells are saturated with water and other substances. Seasoning the lumber means evaporation of the water from the cells, leaving the wood not only considerably lighter, but also much stronger.

Illustration 57 (page 98) shows the following growth features of a tree. The *bark* consists of a thin inner growing layer and a thick outer dead layer that serves to protect the living cells

Table 5. *Joist and Rafter Span Tables*

Rough-Sawed Lumber
Uniformly Distributed Load 25 psf
Dead Load 5 psf, Live Load 20 psf

b × d	S	o.c.	f = 900 psi	1,000	1,100	1,200	1,300	1,400	1,500
2 × 4	5.33	12	11–4	11–11	12–6	13–1	13–7	14–1	14–7
		16	9–10	10–4	10–10	11–4	11–9	12–3	12–8
		24	8–0	8–5	8–10	9–3	9–7	10–0	10–4
		48	5–8	6–0	6–3	6–6	6–10	7–1	7–4
2 × 6	12.0	12	17–0	17–11	18–9	19–7	20–5	21–2	21–4
		16	14–8	15–6	16–3	17–0	17–8	18–4	19–0
		24	12–0	12–8	13–3	13–10	14–5	15–0	15–6
		48	8–6	8–11	9–5	9–10	10–2	10–7	10–11
2 × 8	21.3	12	22–7	23–10	>24	—	—	—	—
		16	19–7	20–8	21–8	22–8	23–6	>24	—
		24	16–0	16–10	17–8	18–6	19–3	20–0	20–8
		48	11–4	11–11	12–6	13–1	13–7	14–1	14–7
2 × 10	33.3	12	>24	—	—	—	—	—	—
		16	>24	—	—	—	—	—	—
		24	20–0	21–1	22–1	23–1	24–0	>24	—
		48	14–2	14–11	15–8	16–4	17–0	17–8	18–3
2 × 12	48.0	12	>24	—	—	—	—	—	—
		16	>24	—	—	—	—	—	—
		24	24–0	>24	—	—	—	—	—
		48	17–0	17–11	18–9	19–7	20–5	21–2	21–11
3 × 6	18.0	16	18–0	19–0	19–11	20–9	21–8	22–5	23–3
		24	14–8	15–6	16–3	17–0	17–8	18–4	19–0
		48	10–5	10–11	11–6	12–0	12–6	13–0	13–5
3 × 8	32.0	16	24–0	>24	—	—	—	—	—
		24	19–7	20–8	21–8	22–8	23–6	>24	—
		48	13–10	14–7	15–4	16–0	16–8	17–3	17–11
3 × 10	50.0	16	>24	—	—	—	—	—	—
		24	>24	—	—	—	—	—	—
		48	17–4	18–3	19–2	20–0	20–10	21–7	22–4
3 × 12	22.0	16	>24	—	—	—	—	—	—
		24	>24	—	—	—	—	—	—
		48	20–9	21–11	23–0	24–0	>24	—	—
4 × 6	24.0	24	17–0	17–11	18–9	19–7	20–5	21–2	21–11
		48	12–0	12–8	13–3	13–10	14–5	15–0	15–6
4 × 8	42.7	24	22–8	23–10	>24	—	—	—	—
		48	16–0	16–10	17–8	18–6	19–3	20–0	20–8
4 × 10	66.7	24	>24	—	—	—	—	—	—
		48	20–0	21–1	22–1	23–1	24–0	>24	—
4 × 12	96.0	24	>24	—	—	—	—	—	—
		48	24–0	>24	—	—	—	—	—
6 × 6	36.0	48	20–9	15–6	16–3	17–0	17–8	18–4	19–0
6 × 8	64.0	48	19–7	20–8	21–8	22–8	23–6	>24	—
6 × 10	100	48	>24	—	—	—	—	—	—
6 × 12	144	48	>24	—	—	—	—	—	—
8 × 8	85.3	48	22–7	23–10	>24	—	—	—	—
8 × 10	133	48	>24	—	—	—	—	—	—
8 × 12	192	48	>24	—	—	—	—	—	—

Verify by Case 1 formula

Rough-Sawed Lumber
Uniformly Distributed Load 35 psf
Dead Load 5 psf, Live Load 30 psf

b × d	S	o.c.	f = 900 psi	1,000	1,100	1,200	1,300	1,400	1,500
2 × 4	5.33	12	9–6	10–1	10–7	11–0	11–6	11–11	12–4
		16	8–3	8–9	9–2	9–6	9–11	10–4	10–8
		24	6–9	7–1	7–6	7–10	8–1	8–5	8–9
		48	4–9	5–0	5–3	5–6	5–9	6–0	6–2
2 × 6	12.0	12	14–4	15–1	15–10	16–7	17–3	17–11	18–6
		16	12–5	13–1	13–9	14–4	14–11	15–6	16–0
		24	10–2	10–8	11–3	11–9	12–2	12–8	13–1
		48	7–2	7–7	7–11	8–3	8–7	8–11	9–3
2 × 8	21.3	12	19–1	20–2	21–2	22–1	23–0	23–11	—
		16	16–7	17–6	18–4	19–2	19–11	20–8	21–4
		24	13–6	14–3	14–11	15–7	16–3	16–10	17–5
		48	9–6	10–1	10–7	11–0	11–6	11–11	12–4
2 × 10	33.3	12	23–11	> 24	—	—	—	—	—
		16	20–9	21–10	22–11	23–11	> 24	—	—
		24	16–11	17–10	18–8	19–6	20–4	21–1	21–10
		48	12–0	12–7	13–3	13–10	14–4	14–11	15–5
2 × 12	48.0	12	> 24	—	—	—	—	—	—
		16	> 24	—	—	—	—	—	—
		24	20–3	21–4	22–5	23–5	> 24	—	—
		48	14–4	15–1	15–10	16–7	17–3	17–11	18–6
3 × 6	18.0	16	15–3	16–0	16–10	17–7	18–3	19–0	19–8
		24	12–5	13–1	13–9	14–4	14–11	15–6	16–0
		48	8–9	9–3	9–9	10–2	10–6	11–0	11–4
3 × 8	32.0	16	20–3	21–4	22–5	23–5	> 24	—	—
		24	16–7	17–6	18–4	19–1	19–11	20–8	21–4
		48	11–9	12–4	12–11	13–6	14–1	14–7	15–1
3 × 10	50.0	16	> 24	—	—	—	—	—	—
		24	20–8	21–10	22–11	23–11	> 24	—	—
		48	14–8	15–5	16–2	16–11	17–7	18–3	18–11
3 × 12	72.0	16	> 24	—	—	—	—	—	—
		24	> 24	—	—	—	—	—	—
		48	17–7	18–6	19–5	20–3	21–1	21–11	22–8
4 × 6	24.0	24	14–4	15–1	15–10	16–7	17–3	17–11	18–6
		48	10–2	10–8	11–3	11–9	12–2	12–8	13–1
4 × 8	42.7	24	19–1	20–2	21–2	22–1	23–0	23–10	> 24
		48	13–6	14–3	15–0	15–7	16–3	16–10	17–6
4 × 10	66.7	24	23–11	> 24	—	—	—	—	—
		48	16–11	17–10	18–8	19–6	20–4	21–1	21–10
4 × 12	96.0	24	—	—	—	—	—	—	—
		48	20–3	21–4	22–5	23–5	> 24	—	—
6 × 6	36.0	48	12–5	13–1	13–9	14–4	14–11	15–6	16–0
6 × 8	64.0	48	16–7	17–5	18–4	19–1	19–11	20–8	21–4
6 × 10	100	48	20–8	21–10	22–11	23–11	> 24	—	—
6 × 12	144	48	> 24	—	—	—	—	—	—
8 × 8	85.3	48	19–1	20–2	21–2	22–1	23–0	23–10	> 24
8 × 10	133	48	23–11	> 24	—	—	—	—	—
8 × 12	192	48	—	—	—	—	—	—	—

Verify by Case 1 formula

Rough-Sawed Lumber
Uniformly Distributed Load 37.5 psf
Dead Load 7.5 psf, Live Load 30 psf

$b \times d$	S	o.c.	f = 900 psi	1,000	1,100	1,200	1,300	1,400	1,500
2 × 4	5.33	12	9–3	9–9	10–3	10–8	11–1	11–6	12–1
		16	8–0	8–5	8–10	9–3	9–7	10–0	10–4
		24	6–6	6–11	7–3	7–6	7–10	8–2	8–5
		48	4–8	4–10	5–1	5–4	5–6	5–9	6–0
2 × 6	12.0	12	13–10	14–7	15–4	16–0	16–8	17–3	17–11
		16	12–0	12–8	13–3	13–10	14–5	15–0	15–6
		24	9–10	10–4	10–10	11–4	11–9	12–3	12–8
		48	6–11	7–4	7–8	8–0	8–4	8–8	8–11
2 × 8	21.3	12	18–6	19–6	20–5	21–4	22–2	23–0	12–10
		16	16–0	16–10	17–8	18–6	19–3	20–0	20–8
		24	13–1	13–9	14–5	15–1	15–8	16–4	16–10
		48	9–3	9–9	10–3	10–8	11–1	11–6	11–11
2 × 10	33.3	12	23–1	> 24	—	—	—	—	—
		16	20–0	21–1	22–1	23–1	24–0	> 24	—
		24	16–4	17–3	18–1	18–10	19–8	20–4	21–1
		48	11–6	12–2	12–9	13–4	13–10	14–5	14–11
2 × 12	48.0	12	> 24	—	—	—	—	—	—
		16	24–0	> 24	—	—	—	—	—
		24	19–7	20–8	21–8	22–8	23–6	> 24	—
		48	13–10	14–7	15–4	16–0	16–8	17–3	17–10
3 × 6	18.0	16	14–8	15–6	16–3	17–0	17–8	18–4	19–0
		24	12–0	12–8	13–3	13–10	14–5	15–0	15–6
		48	8–6	8–11	9–5	9–10	10–3	10–7	11–0
3 × 8	32.0	16	19–7	20–8	21–8	22–8	23–6	> 24	—
		24	16–0	16–10	17–8	18–6	19–3	20–0	20–8
		48	11–4	11–11	12–6	13–1	13–7	14–1	14–7
3 × 10	50.0	16	> 24	—	—	—	—	—	—
		24	20–0	21–1	22–1	23–1	> 24	—	—
		48	14–2	14–11	15–8	16–4	17–0	17–8	18–3
3 × 12	72.0	16	> 24	—	—	—	—	—	—
		24	> 24	—	—	—	—	—	—
		48	17–0	17–10	18–9	19–7	20–5	21–2	21–11
4 × 6	24.0	24	13–10	14–7	15–4	16–0	16–8	17–3	17–10
		48	9–9	10–4	10–10	11–4	11–9	12–3	12–8
4 × 8	42.7	24	18–6	19–6	20–5	21–4	22–3	23–0	23–10
		48	13–1	13–9	14–5	15–1	15–8	16–3	16–10
4 × 10	66.7	24	> 24	—	—	—	—	—	—
		48	16–4	17–3	18–1	18–10	19–8	20–4	21–1
4 × 12	96.0	24	> 24	—	—	—	—	—	—
		48	19–7	20–8	21–8	22–8	23–6	> 24	—
6 × 6	36.0	48	12–0	12–8	13–3	13–10	14–5	15–0	15–6
6 × 8	64.0	48	16–0	16–10	17–8	18–6	19–3	20–0	20–8
6 × 10	100	48	20–0	21–1	22–1	23–1	24–0	> 24	—
6 × 12	144	48	24–0	> 24	—	—	—	—	—
8 × 8	85.3	48	18–6	19–6	20–5	21–4	22–3	23–0	23–10
8 × 10	133	48	23–1	24–0	> 24	—	—	—	—
8 × 12	192	48	> 24	—	—	—	—	—	—

Verify by Case 1 formula

b × d	S	o.c.	f = 900 psi	1,000	1,100	1,200	1,300	1,400	1,500
2 × 4	5.33	12	8–5	8–10	9–4	9–9	10–2	10–6	10–11
		16	7–4	7–9	8–1	8–5	8–9	9–1	9–5
		24	6–0	6–3	6–7	6–11	7–2	7–5	7–8
		48	4–3	4–5	4–8	4–10	5–1	5–3	5–5
2 × 6	12.0	12	12–8	13–4	14–0	14–7	15–3	15–9	16–4
		16	11–0	11–6	12–1	12–8	13–2	13–8	14–2
		24	8–11	9–5	9–11	10–4	10–9	11–2	11–6
		48	6–4	6–8	7–0	7–4	7–7	7–11	8–2
2 × 8	21.3	12	16–10	17–9	18–8	19–6	20–3	21–0	21–9
		16	14–7	15–5	16–2	16–10	17–6	18–3	18–10
		24	11–11	12–7	13–2	13–9	14–4	14–10	15–5
		48	8–5	8–11	9–4	9–9	10–2	10–6	10–11
2 × 10	33.3	12	21–1	22–3	23–4	>24	—	—	—
		16	18–3	19–3	20–2	21–1	21–11	22–9	23–7
		24	14–11	15–9	16–6	17–3	17–11	18–7	19–3
		48	10–6	11–1	11–8	12–2	12–8	13–2	13–7
2 × 12	48.0	12	>24	—	—	—	—	—	—
		16	21–11	23–1	>24	—	—	—	—
		24	17–11	18–10	19–9	20–8	21–6	22–4	23–1
		48	12–8	13–4	14–0	14–7	15–2	15–9	16–4
3 × 6	18.0	16	13–5	14–2	14–10	15–6	16–2	16–9	17–4
		24	11–0	11–6	12–1	12–8	13–2	13–8	14–2
		48	7–9	8–2	8–7	8–11	9–4	9–8	10–0
3 × 8	32.0	16	17–10	18–10	19–9	20–8	21–6	22–4	23–1
		24	14–7	15–5	16–2	16–10	17–6	18–3	18–10
		48	10–4	10–11	11–5	11–11	12–5	12–11	13–4
3 × 10	50.0	16	22–4	23–7	>24	—	—	—	—
		24	18–3	19–3	20–2	21–1	21–11	22–9	23–7
		48	12–11	13–7	14–3	14–11	15–6	16–1	16–8
3 × 12	72.0	16	>24	—	—	—	—	—	—
		24	21–11	23–1	>24	—	—	—	—
		48	15–6	16–4	17–2	17–11	18–8	19–4	20–6
4 × 6	24.0	24	12–8	13–4	14–0	14–7	15–2	15–9	16–4
		48	8–11	9–5	9–11	10–4	10–9	11–2	11–6
4 × 8	42.7	24	16–10	17–9	18–8	19–6	20–3	21–0	21–9
		48	11–11	12–7	13–2	13–9	14–4	14–11	15–5
4 × 10	66.7	24	21–11	22–3	23–3	>24	—	—	—
		48	14–11	15–9	16–6	17–3	17–11	18–7	19–3
4 × 12	96.0	24	>24	—	—	—	—	—	—
		48	17–11	18–10	19–9	20–8	21–6	22–4	23–1
6 × 6	36.0	48	11–0	11–6	12–1	12–8	13–2	13–8	14–2
6 × 8	64.0	48	14–7	15–4	16–2	16–10	17–6	18–3	18–10
6 × 10	100	48	18–3	19–3	20–2	21–1	21–11	22–9	23–7
6 × 12	144	48	21–11	23–1	>24	—	—	—	—
8 × 8	85.3	48	16–10	17–9	18–8	19–6	20–3	21–0	21–9
8 × 10	133	48	21–1	22–3	23–4	>24	—	—	—
8 × 12	192	48	>24	—	—	—	—	—	—

Verify by Case 1 formula

Rough-Sawed Lumber
Uniformly Distributed Load 47.5 psf
Dead Load 7.5 psf, Live Load 40 psf

b × d	S	o.c.	f = 900 psi	1,000	1,100	1,200	1,300	1,400	1,500
2 × 4	5.33	12	8-2	8-8	9-1	9-5	9-10	10-3	10-7
		16	7-1	7-6	7-10	8-2	8-6	8-10	9-2
		24	5-10	6-1	6-5	6-8	7-0	7-3	7-6
		48	4-1	4-4	4-6	4-9	4-11	5-1	5-4
2 × 6	12.0	12	12-4	13-6	13-7	14-3	14-9	15-4	15-11
		16	10-9	11-3	11-9	12-3	12-10	13-4	13-9
		24	8-8	9-2	9-8	10-0	10-6	10-10	11-3
		48	6-2	6-6	6-10	7-1	7-5	7-8	7-11
2 × 8	21.3	12	16-5	17-4	18-2	19-0	19-9	20-6	21-2
		16	14-3	15-6	15-9	16-5	17-1	17-9	18-4
		24	11-7	12-3	12-10	13-5	13-11	14-6	15-6
		48	8-3	8-8	9-1	9-6	9-10	10-3	10-7
2 × 10	33.3	12	20-6	21-8	22-8	23-8	> 24	—	—
		16	17-9	18-9	19-8	20-6	21-4	22-2	22-11
		24	14-6	15-4	16-0	16-9	17-5	18-1	18-9
		48	10-3	10-10	11-4	11-10	12-4	12-10	13-3
2 × 12	48.0	12	> 24	—	—	—	—	—	—
		16	21-4	22-6	23-7	> 24	—	—	—
		24	17-5	18-4	19-3	20-1	20-11	21-9	22-6
		48	12-4	13-0	13-7	14-3	14-10	15-4	15-11
3 × 6	18.0	16	13-1	13-9	14-5	15-1	15-8	16-3	16-10
		24	10-8	11-3	11-9	12-4	12-10	13-4	13-9
		48	7-6	7-11	8-4	8-8	9-1	9-5	9-9
3 × 8	32.0	16	17-5	18-4	19-3	20-1	20-11	21-9	22-6
		24	14-3	15-0	15-9	16-5	17-1	17-9	18-4
		48	10-0	10-7	11-1	11-7	12-1	12-6	13-0
3 × 10	50.0	16	21-9	22-11	24-1	> 24	—	—	—
		24	17-9	18-9	19-8	20-6	21-4	22-2	22-11
		48	12-7	13-3	13-11	14-6	15-1	15-8	16-3
3 × 12	72.0	16	> 24	—	—	—	—	—	—
		24	21-24	22-6	23-7	> 24	—	—	—
		48	15-1	15-11	16-8	17-5	18-1	18-10	19-6
4 × 6	24.0	24	12-4	13-0	13-7	14-3	14-9	15-4	15-11
		48	8-8	9-2	9-8	10-0	10-6	10-10	11-3
4 × 8	42.7	24	16-5	17-4	18-2	19-0	19-9	20-6	21-2
		48	11-7	12-3	12-10	13-5	13-11	14-6	15-0
4 × 10	66.7	24	20-6	21-8	22-8	23-8	> 24	—	—
		48	14-6	15-4	16-0	16-9	17-5	18-1	18-9
4 × 12	96.0	24	> 24	—	—	—	—	—	—
		48	17-5	18-4	19-3	20-1	20-11	21-9	22-6
6 × 6	36.0	48	10-8	11-3	11-9	12-14	12-10	13-4	13-9
6 × 8	64.0	48	14-3	15-0	15-9	16-5	17-1	17-9	18-4
6 × 10	100	48	17-9	18-9	19-8	20-6	21-4	22-2	22-11
6 × 12	144	48	21-4	22-6	23-7	> 24	—	—	—
8 × 8	85.3	48	16-5	17-4	18-2	19-0	19-9	20-6	21-2
8 × 10	133	48	20-6	21-8	22-8	23-8	> 24	—	—
8 × 12	192	48	24-7	> 24	—	—	—	—	—

Verify by Case 1 formula

b × d	S	o.c.	f = 900 psi	1,000	1,100	1,200	1,300	1,400	1,500
2 × 4	3.06	12	8–7	9–0	9–6	9–11	10–4	10–8	11–1
		16	7–5	7–10	8–2	8–7	8–11	9–3	9–7
		24	6–1	6–5	6–8	7–0	7–3	7–7	7–10
		48	4–3	4–6	4–9	4–11	5–2	5–4	5–6
2 × 6	7.56	12	13–6	14–2	14–11	15–7	16–2	16–10	17–5
		16	11–8	12–4	12–11	13–6	14–0	14–7	15–1
		24	9–6	10–0	10–6	11–0	11–5	11–11	12–4
		48	6–9	7–1	7–5	7–9	8–1	8–5	8–8
2 × 8	13.1	12	17–9	18–9	19–8	20–6	21–4	22–2	22–11
		16	15–5	16–3	17–0	17–9	18–6	19–2	19–10
		24	12–7	13–3	13–11	14–6	15–1	15–8	16–3
		48	8–11	9–4	9–10	10–3	10–8	11–1	11–6
2 × 10	21.4	12	22–8	23–11	> 24	—	—	—	—
		16	19–7	20–8	21–8	22–8	23–7	> 24	—
		24	16–0	16–11	17–9	18–6	19–3	20–0	20–8
		48	11–4	11–11	12–6	13–1	13–7	14–2	14–7
2 × 12	31.6	12	> 24	—	—	—	—	—	—
		16	23–10	> 24	—	—	—	—	—
		24	19–6	20–6	21–6	22–6	23–5	> 24	—
		48	13–9	14–6	15–3	15–11	16–6	17–2	17–9
3 × 6	12.6	16	15–1	15–11	16–8	17–4	18–1	18–9	19–5
		24	12–4	13–0	13–7	14–2	14–9	15–4	15–11
		48	8–8	9–2	9–7	10–0	10–5	10–10	11–3
3 × 8	21.9	16	19–10	20–11	21–11	22–11	23–4	> 24	—
		24	16–3	17–1	17–11	18–9	19–6	20–3	20–11
		48	11–6	12–1	12–8	13–3	13–9	14–4	14–10
3 × 10	35.6	16	> 24	—	—	—	—	—	—
		24	20–8	21–10	22–10	23–10	> 24	—	—
		48	14–7	15–5	16–2	16–11	17–7	18–3	18–11
3 × 12	52.7	16	> 24	—	—	—	—	—	—
		24	> 24	—	—	—	—	—	—
		48	17–9	18–9	19–8	20–6	21–5	22–2	23–0
4 × 6	17.6	24	14–6	15–4	16–1	16–9	17–6	18–2	18–9
		48	10–3	10–10	11–4	11–11	12–4	12–10	13–3
4 × 8	30.7	24	19–2	20–2	21–2	22–2	23–1	23–11	> 24
		48	13–7	14–4	15–0	15–8	16–4	16–11	17–6
4 × 10	49.9	24	> 24	—	—	—	—	—	—
		48	17–4	18–3	19–2	20–0	20–10	21–7	22–4
4 × 12	73.8	24	> 24	—	—	—	—	—	—
		48	21–0	22–2	23–3	> 24	—	—	—
6 × 6	27.7	48	12–11	13–7	14–3	14–11	15–6	16–1	16–8
6 × 8	51.6	48	17–7	18–6	19–5	20–3	21–1	21–11	22–8
6 × 10	82.7	48	22–3	23–5	> 24	—	—	—	—
6 × 12	121	48	> 24	—	—	—	—	—	—
8 × 8	70.3	48	20–6	21–7	22–8	23–8	> 24	—	—
8 × 10	113	48	> 24	—	—	—	—	—	—
8 × 12	165	48	> 24	—	—	—	—	—	—

Verify by Case 1 formula

b × d	S	o.c.	f = 900 psi	1,000	1,100	1,200	1,300	1,400	1,500
2 × 4	3.06	12	7–3	7–7	8–0	8–4	8–8	9–0	9–4
		16	6–3	6–7	6–11	7–3	7–6	7–10	8–1
		24	5–1	5–4	5–8	5–11	6–2	6–4	6–7
		48	3–7	3–9	4–0	4–2	4–4	4–6	4–8
2 × 6	7.56	12	11–5	12–0	12–7	13–2	13–8	14–2	14–8
		16	9–10	10–5	10–11	11–5	11–10	12–4	12–9
		24	8–1	8–6	8–11	9–4	9–8	10–0	10–5
		48	5–8	6–0	6–3	6–6	6–10	7–1	7–4
2 × 8	13.1	12	15–0	15–10	16–7	17–4	18–0	18–9	19–5
		16	13–0	13–8	14–4	15–0	15–7	16–3	16–9
		24	10–7	11–2	11–9	12–3	12–9	13–3	13–8
		48	7–6	7–10	8–3	8–8	9–0	9–4	9–8
2 × 10	21.4	12	19–2	20–2	21–2	22–1	23–0	23–11	> 24
		16	16–7	17–6	18–4	19–2	19–11	20–8	21–5
		24	13–6	14–3	15–0	15–8	16–3	16–11	17–6
		48	9–6	10–1	10–7	11–0	11–6	11–11	12–4
2 × 12	31.6	12	23–3	> 24	—	—	—	—	—
		16	20–2	21–3	22–4	23–3	> 24	—	—
		24	16–6	17–4	18–2	19–0	19–10	20–6	21–3
		48	11–7	12–3	12–10	13–5	14–0	14–6	15–0
3 × 6	12.6	16	12–9	13–5	14–1	14–8	15–4	15–10	16–5
		24	10–5	10–11	11–6	12–0	12–6	12–11	13–5
		48	7–4	7–9	8–1	8–6	8–10	9–2	9–6
3 × 8	21.9	16	16–9	17–8	18–6	19–4	20–2	20–11	21–8
		24	13–8	14–5	15–2	15–9	16–5	17–1	17–8
		48	9–8	10–2	10–8	11–2	11–7	12–1	12–6
3 × 10	35.6	16	21–5	22–6	23–8	> 24	—	—	—
		24	17–5	18–5	19–4	20–2	21–0	21–9	22–6
		48	12–4	13–0	13–8	14–3	14–10	15–5	15–11
3 × 12	52.7	16	> 24	—	—	—	—	—	—
		24	21–3	22–5	23–6	> 24	—	—	—
		48	15–0	15–10	16–7	17–4	18–1	18–9	19–5
4 × 6	17.6	24	12–3	12–11	13–7	14–2	14–9	15–4	15–10
		48	8–8	9–2	9–7	10–0	10–5	10–10	11–2
4 × 8	30.7	24	16–2	17–1	17–11	18–8	19–5	20–2	20–11
		48	11–5	12–1	12–8	13–2	13–9	14–3	14–9
4 × 10	49.9	24	20–8	21–9	22–10	23–10	> 24	—	—
		48	14–7	15–4	16–2	16–16	17–6	18–3	18–10
4 × 12	73.8	24	> 24	—	—	—	—	—	—
		48	17–9	18–9	19–8	20–6	21–4	22–2	22–11
6 × 6	27.7	48	10–10	11–5	12–0	12–7	13–1	13–7	14–0
6 × 8	51.6	48	14–10	15–8	16–5	17–2	17–11	18–6	19–2
6 × 10	82.7	48	18–10	19–10	20–9	21–8	22–7	23–5	> 24
6 × 12	121	48	22–9	24–0	> 24	—	—	—	—
8 × 8	70.3	48	17–4	18–3	19–2	20–0	20–10	21–7	22–4
8 × 10	113	48	22–0	23–2	> 24	—	—	—	—
8 × 12	165	48	> 24	—	—	—	—	—	—

Verify by Case 1 formula

$b \times d$	S	o.c.	$f = 900$ psi	1,000	1,100	1,200	1,300	1,400	1,500
2 × 4	3.06	12	7–1	7–5	7–9	8–2	8–6	8–9	9–1
		16	6–1	6–5	6–9	7–1	7–4	7–7	7–11
		24	5–0	5–3	5–6	5–9	6–0	6–3	6–5
		48	3–6	3–8	3–10	4–0	4–2	4–4	4–6
2 × 6	7.56	12	11–1	11–8	12–3	12–9	13–4	13–10	14–4
		16	9–7	10–1	10–7	11–1	11–6	12–0	12–5
		24	7–10	8–3	8–8	9–1	9–5	9–9	10–1
		48	5–6	5–9	6–1	6–4	6–7	6–10	7–1
2 × 8	13.1	12	14–7	15–5	16–2	16–10	17–7	18–2	18–10
		16	12–8	13–4	14–0	14–7	15–2	15–9	16–4
		24	10–4	10–11	11–5	11–11	12–5	12–10	13–4
		48	7–3	7–7	8–0	8–4	8–8	9–0	9–4
2 × 10	21.4	12	18–7	19–8	20–7	21–6	22–5	23–3	24–0
		16	16–2	17–0	17–10	18–7	19–5	20–1	20–10
		24	13–2	13–11	14–7	15–2	15–10	16–5	17–0
		48	9–3	9–8	10–2	10–7	11–1	11–6	11–11
2 × 12	31.6	12	22–6	23–9	>24	—	—	—	—
		16	19–5	20–6	21–6	22–6	23–5	>24	—
		24	15–10	16–9	17–7	18–4	19–1	19–10	20–6
		48	11–3	11–10	12–5	13–0	13–6	14–0	14–6
3 × 6	12.6	16	12–3	12–11	13–7	14–2	14–9	15–4	15–10
		24	10–0	10–6	11–1	11–7	12–0	12–6	12–11
		48	7–1	7–5	7–10	8–2	8–6	8–10	9–2
3 × 8	21.9	16	16–2	17–1	17–11	18–8	19–5	20–2	20–11
		24	13–2	13–11	14–7	15–3	15–10	16–6	17–1
		48	9–4	9–10	10–4	10–10	11–3	11–8	12–1
3 × 10	35.6	16	20–8	21–9	22–10	23–10	>24	—	—
		24	16–10	17–9	18–8	19–6	20–3	21–0	21–9
		48	11–11	12–6	13–2	13–9	14–4	14–10	15–5
3 × 12	52.7	16	>24	—	—	—	—	—	—
		24	20–6	21–7	22–8	23–8	>24	—	—
		48	14–6	15–3	16–0	16–9	17–5	18–1	18–9
4 × 6	17.6	24	11–10	12–6	13–1	13–8	14–3	14–9	15–4
		48	8–4	8–10	9–3	9–8	10–1	10–5	10–10
4 × 8	30.7	24	15–8	16–5	17–3	18–0	18–10	19–6	20–2
		48	11–1	11–8	12–3	12–9	13–3	13–9	14–3
4 × 10	49.9	24	20–0	21–0	22–1	23–0	24-0	>24	—
		48	14–1	14–10	15–7	16–3	16–11	17–7	18–3
4 × 12	73.8	24	>24	—	—	—	—	—	—
		48	17–2	18–1	19–0	19–10	20–7	21–5	22–2
6 × 6	27.7	48	10–6	11–0	11–7	12–1	12–7	13–0	13–6
6 × 8	51.6	48	14–4	15–1	15–10	16–7	17–3	17–11	18–6
6 × 10	82.7	48	18–2	19–2	20–1	21–0	21–10	22–8	23–5
6 × 12	121	48	22–0	23–2	>24	—	—	—	—
8 × 8	70.3	48	16–9	17–8	18–6	19–4	20–2	20–10	21–7
8 × 10	113	48	21–3	22–4	23–5	>24	—	—	—
8 × 12	165	48	>24	—	—	—	—	—	—

Verify by Case 1 formula

Dimension S4S Lumber
Uniformly Distributed Load 45 psf
Dead Load 5 psf, Live Load 40 psf

b × d	S	o.c.	f = 900 psi	1,000	1,100	1,200	1,300	1,400	1,500
2 × 4	3.06	12	6-5	6-9	7-1	7-5	7-8	8-0	8-3
		16	5-6	5-10	6-1	6-5	6-8	6-11	7-2
		24	4-6	4-9	5-0	5-3	5-5	5-8	5-10
		48	3-2	3-4	3-6	3-8	3-10	4-0	4-1
2 × 6	7.56	12	10-0	10-7	11-1	11-7	12-1	12-6	13-0
		16	8-8	9-2	9-7	10-0	10-5	10-10	11-3
		24	7-1	7-6	7-10	8-2	8-6	8-10	9-2
		48	5-0	5-3	5-6	5-9	6-0	6-3	6-5
2 × 8	13.1	12	13-3	13-11	14-8	15-3	15-11	16-6	17-1
		16	11-6	12-1	12-8	13-3	13-9	14-4	14-10
		24	9-4	9-10	10-4	10-10	11-3	11-8	12-1
		48	6-7	6-11	7-3	7-7	7-11	8-3	8-6
2 × 10	21.4	12	16-11	17-10	18-8	19-6	20-4	21-1	21-10
		16	14-8	15-5	16-2	16-11	17-7	18-3	18-11
		24	11-11	12-7	13-2	13-9	14-4	14-11	15-5
		48	8-5	8-10	9-4	9-9	10-2	10-6	10-10
2 × 12	31.6	12	20-6	21-8	22-8	23-9	>24	—	—
		16	17-9	18-9	19-8	20-6	21-5	22-2	23-0
		24	14-6	15-4	16-1	16-9	17-5	18-1	18-9
		48	10-3	10-10	11-4	11-10	12-4	12-9	13-3
3 × 6	12.6	16	11-3	11-10	12-5	13-0	13-6	14-0	14-6
		24	9-2	9-8	10-2	10-7	11-0	11-5	11-10
		48	6-6	6-10	7-2	7-6	7-9	8-1	8-4
3 × 8	21.9	16	14-9	15-7	16-4	17-1	17-9	18-5	19-1
		24	12-1	12-9	13-4	14-0	14-6	15-1	15-7
		48	8-6	9-0	9-5	9-10	10-3	10-8	11-0
3 × 10	35.6	16	18-10	19-11	20-10	21-9	22-8	23-6	>24
		24	15-5	16-3	17-0	17-9	18-6	19-3	19-11
		48	10-10	11-6	12-0	12-7	13-1	13-7	14-0
3 × 12	52.7	16	23-0	>24	—	—	—	—	—
		24	18-9	19-9	20-9	21-8	22-6	23-5	>24
		48	13-3	13-11	14-8	15-3	15-11	16-6	17-1
4 × 6	17.6	24	10-10	11-5	11-11	12-6	13-0	13-6	14-0
		48	7-8	8-1	8-5	8-10	9-2	9-6	9-10
4 × 8	30.7	24	14-3	15-0	15-9	16-6	17-2	17-9	18-5
		48	10-1	10-7	11-2	11-8	12-1	12-7	13-0
4 × 10	49.9	24	18-3	19-2	20-1	21-0	21-11	22-9	23-6
		48	12-10	13-7	14-3	14-10	15-6	16-1	16-7
4 × 12	73.8	24	22-2	23-4	>24	—	—	—	—
		48	15-8	16-6	17-4	18-1	18-10	19-6	20-3
6 × 6	27.7	48	9-7	10-1	10-7	11-1	11-6	12-1	12-5
6 × 8	51.6	48	13-1	13-9	14-6	15-1	15-9	16-4	16-11
6 × 10	82.7	48	16-7	17-6	18-4	19-2	19-11	20-8	21-5
6 × 12	121	48	20-1	21-2	22-2	23-2	>24	—	—
8 × 8	70.3	48	15-3	16-2	16-11	17-8	18-4	19-1	19-9
8 × 10	113	48	19-4	20-5	21-5	22-4	23-3	>24	—
8 × 12	165	48	23-5	>24	—	—	—	—	—

Verify by Case 1 formula

Dimension S4S Lumber
Uniformly Distributed Load 47.5 psf
Dead Load 7.5 psf, Live Load 40 psf

b × d	S	o.c.	f = 900 psi	1,000	1,100	1,200	1,300	1,400	1,500
2 × 4	3.06	12	6–3	6–7	6–11	7–3	7–6	7–10	8–1
		16	5–5	5–8	6–0	6–3	6–6	6–9	7–0
		24	4–5	4–8	4–11	5–1	5–4	5–6	5–8
		48	3–1	3–3	3–5	3–7	3–9	3–10	4–0
2 × 6	7.56	12	9–10	10–4	10–10	11–4	11–10	12–3	12–8
		16	8–6	9–0	9–5	9–10	10–3	10–7	11–0
		24	6–11	7–4	7–8	8–0	8–4	8–8	9–0
		48	4–10	5–2	5–5	5–7	5–10	6–1	6–3
2 × 8	13.1	12	12–11	13–8	14–4	14–11	15–7	16–2	16–9
		16	11–3	11–10	12–5	12–11	13–6	14–0	14–6
		24	9–2	9–8	10–2	10–7	11–0	11–5	11–10
		48	6–5	6–9	7–1	7–5	7–9	8–0	8–3
2 × 10	21.4	12	16–6	17–5	18–3	19–1	19–10	20–7	21–4
		16	14–4	15–1	15–10	16–6	17–2	17–10	18–6
		24	11–8	12–4	12–11	13–6	14–1	14–7	15–1
		48	8–2	8–8	9–1	9–5	9–10	10–3	10–7
2 × 12	31.6	12	20–0	21–1	22–1	23–1	24–0	> 24	—
		16	17–4	18–3	19–2	20–0	20–9	21–7	22–4
		24	14–2	14–11	15–7	16–4	16–11	17–7	18–3
		48	10–0	10–6	11–0	11–6	12–0	12–5	12–10
3 × 6	12.6	16	10–11	11–6	12–1	12–7	13–2	13–7	14–1
		24	8–11	9–5	9–10	10–3	10–8	11–1	11–6
		48	6–3	6–7	6–11	7–3	7–7	7–10	8–2
3 × 8	21.9	16	14–5	15–2	15–11	16–7	17–4	17–11	18–7
		24	11–9	12–4	13–0	13–7	14–2	14–8	15–2
		48	8–3	8–9	9–2	9–7	10–0	10–4	10–8
3 × 10	35.6	16	18–4	19–4	20–4	21–3	22–1	22–11	23–8
		24	15–0	15–9	16–7	17–4	18–0	18–8	19–4
		48	10–7	11–2	11–8	12–3	12–9	13–2	13–8
3 × 12	52.7	16	22–4	23–6	24–0	> 24	—	—	—
		24	18–3	19–3	20–2	21–1	21–11	22–9	23–6
		48	12–10	13–7	14–3	14–10	15–6	16–1	16–8
4 × 6	17.6	24	10–6	11–1	11–8	12–2	12–8	13–2	13–7
		48	7–5	7–11	8–3	8–7	8–11	9–3	9–7
4 × 8	30.7	24	13–11	14–8	15–4	16–0	16–8	17–4	17–11
		48	9–10	10–4	10–10	11–4	11–10	12–3	12–8
4 × 10	49.9	24	17–9	18–8	19–7	20–6	21–4	22–1	22–11
		48	12–6	13–2	13–10	14–6	15–1	15–7	16–2
4 × 12	73.8	24	21–7	22–9	23–10	> 24	—	—	—
		48	15–3	16–1	16–10	17–7	18–4	19–0	19–8
6 × 6	27.7	48	9–4	9–10	10–4	10–9	11–2	11–8	12–1
6 × 8	51.6	48	12–9	13–5	14–1	14–8	15–4	15–10	16–5
6 × 10	82.7	48	16–2	17–0	17–10	18–8	19–5	20–2	20–10
6 × 12	121	48	19–6	20–7	21–7	22–7	23–6	> 24	—
8 × 8	70.3	48	14–10	15–8	16–5	17–2	17–10	18–7	19–3
8 × 10	113	48	18–10	19–10	20–10	21–9	22–8	23–6	> 24
8 × 12	165	48	22–10	> 24	—	—	—	—	—

Verify by Case 1 formula

	b × d	I, in.⁴	o.c.	E = 0.8 × 10⁶	1.0 × 10⁶	1.2 × 10⁶	1.4 × 10⁶	1.6 × 10⁶
Rough Lumber	2 × 6	36.0	12	9–10	10–7	11–3	11–10	12–4
			16	8–11	9–7	10–3	10–9	11–3
			24	7–10	8–5	8–11	9–5	9–10
	2 × 8	85.3	12	13–1	14–1	15–0	15–7	16–6
			16	11–10	12–10	13–7	14–4	15–0
			24	10–5	11–2	11–11	12–6	13–1
	2 × 10	166.7	12	16–4	17–8	18–9	19–9	20–7
			16	14–10	16–0	17–0	17–11	18–9
			24	13–0	14–0	14–10	15–8	16–4
	2 × 12	288	12	19–8	21–2	22–6	23–8	> 24
			16	17–10	19–3	20–5	21–6	22–6
			24	15–7	16–9	17–10	18–9	19–8
S4S Lumber	2 × 6 (1-1/2 × 5-1/2)	20.8	12	8–9	9–6	10–1	10–7	11–1
			16	8–0	8–8	9–1	9–8	10–1
			24	7–0	7–6	8–0	8–5	8–9
	2 × 8 (1-1/2 × 7-1/4)	47.6	12	11–8	12–6	13–3	14–0	14–8
			16	10–6	11–4	12–1	12–8	13–3
			24	9–3	9–11	10–6	11–1	11–8
	2 × 10 (1-1/2 × 9-1/4)	98.9	12	14–10	15–11	17–0	17–10	18–8
			16	13–5	14–6	15–5	16–3	17–0
			24	11–9	12–8	13–5	14–2	14–10
	2 × 12 (1-1/2 × 11-1/4)	178	12	18–0	19–5	20–7	21–9	22–8
			16	16–4	17–7	18–9	19–9	20–7
			24	14–4	15–5	16–4	17–3	18–0

beneath. The *cambium* is a very thin layer where new wood cells are grown. The *sapwood,* usually light in color, consists partly of dead and partly of still living cells and carries sap up the trunk to the leaves. The *heartwood,* usually darker in color, is old sapwood that no longer functions to carry or store sap. Heartwood is also usually heavier, stronger, and more resistant to decay than sapwood due to chemical changes that have taken place.

The growth rate of new cells varies with the seasons, often resulting in clearly identifiable *growth rings.* The portion of the annual ring originating early in the year is termed the *earlywood* and the portion from later in the season, the *latewood.* Latewood cells are more dense and therefore stronger. The proportion of latewood to earlywood determines the clear specimen strength of the wood of any one species.

Because the long cells are parallel to the trunk (we call the direction of the cells the "grain"), wood is typically ten times stronger along the grain than perpendicular to the grain. The strength of the cell bonds perpendicular to the grain determines the horizontal shear strength. If the bond between growth rings is so weak that the growth rings actually separate under the stresses of drying, the wood is said to have "shake."

Chemicals in the wood, not actually part of the wood structure, are the *extractives,* determining to a large part the color, decay resistance, and odor of the wood.

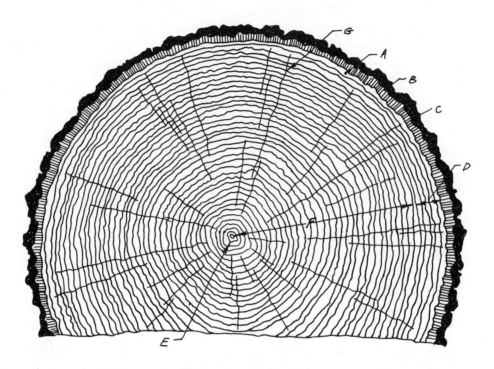

57

Cross section of a tree trunk: A — cambium layer (microscopic) is inside inner bark and forms wood and bark cells. B — inner bark is moist, soft, and contains living tissue. Carries prepared food from leaves to all growing parts of tree. C — outer bark containing corky layers is composed of dry dead tissue. Gives general protection against external injuries. Inner and outer bark are separated by a bark cambium. D — sapwood, which contains both living and dead tissues, is the light-colored wood beneath the bark. Carries sap from roots to leaves. E — heartwood (inactive) is formed by a gradual change in the sapwood. F — pith is the soft tissue about which the first wood growth takes place in the newly formed twigs. G — wood rays connect the various layers from pith to bark for storage and transfer of food.

How Wood Is Sawed

A piece of lumber can be sawed from a log in two ways: primarily tangent to the growth rings (plainsawed or slash-sawed) or primarily perpendicular to the rings (quartersawed or edge-grained). To avoid waste, all of the cuts must be either parallel or perpendicular to the first cut, thereby resulting in most boards being neither strictly perpendicular nor parallel to the rings. The term that comes closest is used. If all of the cuts are made parallel to the first, about half of the boards will be plainsawed and half quartersawed. By rotating the log a quarter turn when necessary, all of the boards can be quartersawed. Quartersawing is rare,

especially in small mills, because it requires more time and therefore, expense.

The relative merits of the two methods are:

Plainsawed
(1) Annual ring patterns are more beautiful, especially in pine.
(2) It costs less.
(3) Knots are round and small compared to the spike knots of quartersawed boards.
(4) The round knots affect the strength of the board less.

Quartersawed
(1) Weathers better, shrinks less in width, twists and cups less, checks and splits less.
(2) Holds paint better.
(3) Wears more evenly as a floor.

Seasoning Wood

Seasoning wood consists of drying the wood from the green state to its equilibrium moisture content in a controlled manner. Large mills force rapid drying of the wood in ovens called kilns. The main advantage of the kiln is speed, taking a week to achieve a state requiring months naturally. A second, but small, advantage is control of the drying process, resulting in fewer seasoning-associated defects. A large millowner confessed to me, however, that *his* only reason for kiln drying was to reduce shipping weight and therefore cost.

Proper seasoning can prove a large factor in achieving the desired qualities in your lumber. If you are out to save money, you have no choice but to use the air-drying method. To assure that no seasoning defects occur, have your lumber delivered immediately after sawing. Getting delivery in early spring or anytime when the average daily temperature is below 50° F will help, since defects will not develop in low temperatures.

Defects that may develop by improper seasoning include:

(1) Checks (splits) caused by the stresses of unequal shrinkage. The ends of boards are particularly susceptible to "end checking" because water is evaporating from the end more rapidly than from the interior. Direct sunlight on the boards' ends aggravates the situation.

(2) Cupping, warping, and twisting — actually the natural reaction of drying and shrinking wood.

(3) Loosening of knots caused by more rapid drying of surrounding wood than of the knot.

(4) Green or black mold, due to the growth of fungi under moist conditions at a temperature over 50° F.

(5) Sticker burn, due to mold and unequal chemical reaction of extractives under "stickers."

Most of the defects are avoided by stacking (or "sticking") the green wood properly within a few days of sawing. Illustration 58 shows a properly built lumber pile. First, select a well-drained site devoid of vegetation. Next, establish a proper foundation of three parallel and level timbers that will bear the weight and keep the bottommost lumber several inches clear of the ground. When you received your green lumber you should also have obtained a sufficient number of dry (not green!) stickers, which are the uneven, bark-covered edges of the boards. The length of the stickers establishes the width of the pile. Lay the first course of boards with the adjacent edges separated by an inch. Then directly above the three foundation timbers lay three stickers. Repeat, alternating layers of boards and stickers. The stickers should be dry to minimize mold. They must also be a minimum of 1 inch thick or the separation of layers will be insufficient to prevent mold. If the boards show beginning signs of mold when received, lay them out individually in the sun for a few days to kill the fungus before sticking.

The weight of the layers above acts as a giant press, preventing warp, cup, and twist. When the wood dries, it "sets" and will remain in this shape until moisture conditions change again. Therefore, high, narrow piles are better than low, wide ones.

The sun produces color changes in the wood. Pine in direct sunlight and rain typically progresses from white to honey in a week, to a faun color in a month, and to gray in a season. If the edges and ends of the piles receive more sunlight you will get color banding. Stick burn is sometimes thereby reversed, producing a light band. I usually pile lumber with an old corrugated tin roof cover for six to eight weeks and then set the boards out individually until

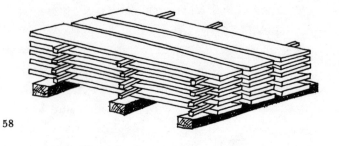

58

the desired color is achieved. Even better is an open drying shed. Wood, once dried, can readily absorb moisture and swell. Beware, therefore, of "kiln-dried" lumber that has sat outdoors at the lumberyard in a solid pile for months.

Shrinkage

In drying from the green to the final equilibrium moisture content of usage, lumber shrinks. Table 6 gives the maximum shrinkage tangential and perpendicular to the growth rings for a few common species.

From the table it is obvious why our ancestors sheathed their houses and barns with pine and cedar and made their implement handles of ash. The percent shrinkage is a pretty good indication of the trouble you'll have with checking and cupping of exposed wood.

Illustration 59 shows how wood typically deforms in shrinking. If lumber is sawed dry and then swells, the deformation will be just the opposite. From the illustration it is seen that wall siding should be applied *heart side out* and floor boards *heart side up*. While sheathing and flooring chant to yourself, "Heart side up, heart side out." Not only will the cupping be less objectionable, but any resulting cracks will occur on the unexposed side and edge nails will not be withdrawn. As a rule of thumb, to minimize cupping, the ratio of width to thickness of exterior boards should never exceed eight (a width of 8 inches for rough boards, or 6 inches for 3/4-inch planed boards).

Grading Framing Lumber

Table 4 gave average strength values for common species of wood used in framing. To use these values we must recognize the possible reduction

Table 6. *Maximum Shrinkage of Wood Species from Green to Oven Dry*

Species	Percent Shrinkage	
	Perpendicular to Rings	*Tangential to Rings*
Ash, White	4.9	7.8
Basswood	6.6	9.3
Birch, Paper	6.3	8.6
Hickory, Shagbark	7.0	10.5
Maple, Sugar	4.8	9.9
Oak, White	5.6	10.5
Cedar, Northern White	2.2	4.9
Fir, Douglas	4.8	7.6
Hemlock, Eastern	3.0	6.8
Pine, Eastern White	2.1	6.1
Pine, Southern	4.8	7.0
Spruce, Eastern	4.0	7.0

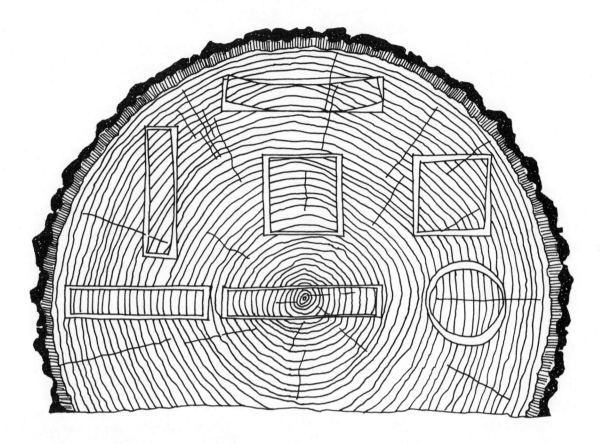

59

Characteristic shrinkage and distortion of flats, squares, and rounds as affected by the direction of the annual rings. Tangential shrinkage is about twice as great as radial.

in strength due to various defects. Commercially graded lumber is downgraded more than necessary because it is assumed that even the professional carpenter will sometimes use a piece in the worst possible way!

Four basic types of defects are most important:

KNOTS

A knot is extremely hard, but the bond between knot and surrounding wood is weak. Therefore a knot has full strength in compression but zero strength in tension. Illustration 60 shows that in a bending beam the top fibers are in compression, the bottom fibers in tension, and the middle fibers in a neutral zone. If we divide the beam into equal parts we can say that a sound knot totally within the top two-thirds has no effect, but a knot within the bottom third effectively reduces the depth of the beam by the knot diameter, i.e., a 2″ × 8″ joist with a 2-inch knot at the bottom edge is no stronger than a perfect 2″ × 6″. If we reverse the joist, however, it may be counted as a 2″ × 8″.

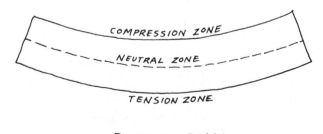

BENDING BEAM

60

ROT

No piece of wood with evidence of rot should be used for framing, as the extent of the rot is difficult to assess.

CHECKS, SPLITS, AND SHAKE

Checks and splits, unless deep and located at an end of a beam in bending, are rarely serious. The split is a sort of stretch mark due to the wood surface shrinking more than the interior, and rarely penetrates to a significant depth. Shake (separation of growth rings), however, can be serious. A visible shake at the end of a board should be assumed to extend the length of the board. Even if it doesn't, the zone of weakness does, effectively reducing the beam to two smaller beams in parallel.

CROSS GRAIN

The angle between the direction of the grain and the long dimension of a beam is termed the cross-grain angle. A cross grain of 1:20 results in a drop of bending strength to 90 percent, 1:10 to 80 percent, and 1:5 to 50 percent. A large cross-grain angle will also probably cause a beam to warp excessively in drying.

Plywood

Plywood consists of veneers of wood peeled from a log, usually each 1/8 inch thick, and bonded with glue. It is only as strong as the wood of which it is made, but it has the distinct advantage of a more nearly uniform strength in all directions. Natural wood has a strength in tension (fiber stress in bending) ratio of about 10:1 (parallel to grain: perpendicular to grain). In plywood it is more nearly 2:1. Therefore, plywood is capable of acting as a membrane and is much less critical in direction of application. Also, its extreme width (48 inches) gives it a distinct advantage over board when used as a flooring or roofing sheathing.

Plywood is described or classified by various characteristics which are generally:

TYPE

Exterior: waterproof glue and high standard for interior plies. *Interior:* not waterproof glue, more interior defects generally. *Interior with EXT Glue:* interior quality plies but waterproof glue.

GROUP

Plywood can be made from any of seventy species of wood varying in strength. The species are assigned one of five groups, with Group I being the strongest (for example, Douglas fir and southern pine veneers.)

GRADE

Appearance Grades
   A — smooth, no knots, minor repairs.
   B — solid surface.
   C — 1-inch knotholes, some splits.
   D — 2-1/2-inch knot holes, large splits.

Engineering Grades

CD – INT – APA, general sheathing.

CD – INT – APA with EXT Glue (CDX), the most common general sheathing.

UNDERLAYMENT – INT – APA, for use under carpeting or linoleum, etc. Also obtainable with EXT Glue.

STRUCTURAL I – CD – INT – APA, for gussets, stressed skin, or box beams. Has EXT Glue. Also available as STRUCTURAL II – CD – INT – APA.

American Plywood Association Certified plywood always carries a grade stamp. Illustration 61 shows a typical grade stamp for 1/2-inch CDX plywood. Most of the symbols are self-explanatory. The key to the mechanical strength is the pair of numbers 32/16. The first number always indicates the maximum allowed rafter spacing when used as roof sheathing; the second number is the maximum joist spacing when used as subflooring. While the spacing is a function both of thickness and veneer species, the following thicknesses *usually* carry the stamp: 3/8″ – 24/0; 1/2″ – 32/16; 5/8″ – 42/20; 3/4″ – 48/24.

A surprising number of people have strong negative feelings about plywood. Some of the reasons expressed are that plywood:

requires inordinate amounts of energy in production;

is unnatural, and representative of all that is wrong with America;

is not organic (same as second reason, apparently);

is cheap and flimsy;

is expensive.

I do not like to look at plywood. I agree, there is something unnatural about its appearance. However, you'll save a lot of time and money and, in some ways, have a better building if you recognize plywood as being a structural

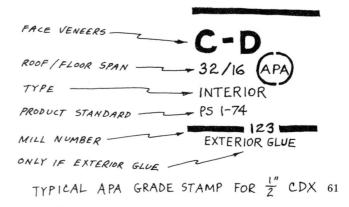

FACE VENEERS ⟶ **C-D**

ROOF/FLOOR SPAN ⟶ 32/16 (APA)

TYPE ⟶ INTERIOR

PRODUCT STANDARD ⟶ PS 1-74

MILL NUMBER ⟶ 123

ONLY IF EXTERIOR GLUE ⟶ EXTERIOR GLUE

TYPICAL APA GRADE STAMP FOR 1/2″ CDX 61

improvement over nature. I can think of ten ways in which plywood makes building either cheaper or easier:

Cost. Plywood wall sheathing costs less per square foot than surfaced boards.

Speed. Plywood sheathing is two to four times faster in application.

Waste. There is virtually no waste using plywood to sheath a building.

Bracing. Nothing is more effective than plywood in making a building rigid.

Shrinkage. Wood shrinks as much as 5 percent across the grain but plywood only 0.2 percent (1/8 inch in 48 inches).

Infiltration. Plywood sheathing blocked at all edges and properly nailed produces a zero infiltration surface.

Strength. The membrane action of plywood makes it nearly indestructible. Imperfections in the plies don't extend through the sheet. The strength of plywood is nearly uniform in all directions.

Vapor Barrier. Plywood with exterior glue is a vapor barrier second only to polyethylene and aluminum foil. Therefore, the vapor barrier in a plywood sub-floor can be eliminated, provided all plywood edges are blocked and nailed.

Precut. Plywood minimizes cutting on the site. It is so square that it may be used as a giant framing square!

Availability. Rough lumber sometimes requires a lead time of eight to twelve weeks to order and season. Plywood is available dry in any quantity, ready to go, in a few days at most.

12

Nails

Before television, before the Super Bowl, before there ever was an American Football League, my brother Chick and I used to be pro football fans. In New York City, our hometown, there was just one place pro football was played: the Polo Grounds, in the Bronx.

Our family home was in Manhattan. On Sunday afternoon (even in those days, reserved for football every autumn), after doing our best not to show impatience at the full-dress Sunday dinner the parents always proclaimed somewhere around 1 P.M., Chick and I would race from the table, out the door, and up Lexington Avenue to the nearest subway station.

We must have been slow learners because, more often than not, we would be uncertain of what train to take to the Polo Grounds. Chick was always too embarrassed to ask; he was so put off by the notion of going public with the admission that he didn't know his way around that he also refused to let me ask.

He'd choose a train, on we'd lunge, and then we'd wait out the long ride uptown, hoping to hear the conductor call the Polo Grounds stop. I don't think we ever guessed right. The train would stop quite emphatically for its last stop, the conductor would yell, "Woodlawn, Woodlawn!" and Chick and I would leave the car along with a handful of somber folks dressed in dark clothes and carrying bouquets to place on one of the tens of thousands of graves at the city cemetery in Woodlawn.

To get the proper train for the Polo Grounds, we'd have to travel all the way back to Eighty-sixth Street. By that time we'd surely missed the first half.

If Chick and I had ever realized how much football it cost us to act wise, we might never have been so reluctant to go public with our innocent ignorance.

Don't be similarly shy about asking for information on nails.

Nails are like the Polo Grounds. Most everyone believes they already know what they need to about both. Like myself and good old Chick, you probably never ask about nails. You just walk into the hardware store or lumberyard and ask for nails; perhaps you say "small ones" or "big ones" but usually it's merely "nails."

If the clerk, trying to be helpful, asks, "Common or finishing?" or "Galvanized or bright?" you probably answer "Common." And that's that. I wonder sometimes how many pounds of unused, unusable common nails there are stashed collectively in the garages, basements, and workshops of the nation of fix-it-yourself handymen.

Like the conductors on the IRT, who knew for sure how to get to the Polo Grounds (if only they'd been asked), the clerks in almost every hardware store and lumberyard in the country know which nails are best for what job and purpose. Their answers to 99 percent of your nail questions will be brief, direct, and eminently understandable. And unless you are planning to build a shelter using wooden pegs for fastenings, you will have to use nails.

Avoid your Woodlawns. Ask for directions. If the thought panics you, read the following. If you are a bit brighter than Chick and I, you'll know the answers when you finish.

(JNC)

13

Nails —
Some Details

The holding power of a common nail is the friction of compressed wood against the nail surface. Illustration 63 shows how the wood

63

fibers are compressed and pushed down out of the way when a nail is driven.

Proper nailing consists of maximizing the frictional force against withdrawal, which is a balance between effects: 1) The area of the nail surface is proportional to the nail length and diameter. 2) The force per area, or pressure, increases with diameter up to the point of

splitting when the pressure dramatically drops. Therefore, the use of too large a nail can be as bad or worse than too small a nail.

The common nail depends entirely upon friction for its action. Other types of nails have been developed that increase resistance to withdrawal by other means.

The Size of Nails

The size of a nail is indicated by *d*, the "pennyweight." The length of all nails of the same penny size is the same, regardless of the head or shank type. However, the diameter may change, resulting in a different number per pound. Table 7 lists nail sizes and number per pound.

Table 7. *Nail Sizes and Numbers per Pound of Common and Box Nails*

Size	Length (in.)	Common Nails		Box Nails	
		Diam. (in.)	#/lb.	Diam. (in.)	#/lb.
4d	1-1/2	.102	316	.083	473
6d	2	.115	181	.102	236
8d	2-1/2	.131	106	.115	145
10d	3	.148	69	.127	94
12d	3-1/4	.148	63	.127	88
16d	3-1/2	.165	49	.134	71
20d	4	.203	31	.148	52
30d	4-1/2	.220	24	.148	46

Types of Nails

Illustration 64 shows the useful house-building nails. *Common* and *box* nails are for normal building construction, particularly framing. Box nails are of smaller diameter and thus have less tendency to split wood. They are particularly useful in nailing boards.

Casing, or *finish,* nails are of small diameter and also have small heads that may be countersunk. They are used where the nailhead shouldn't show. Note that the holding power of the small head is much reduced.

Scaffold nails have a double head. They are driven to the first head. The second head allows later removal with minimal marring of the wood. They are used in temporary application.

Roofing nails have an extremely large, flat head and a barbed shank. This gives them large withdrawal resistance in a thin roof sheathing such as plywood and prevents tearing of the soft roofing material.

Drywall nails have a thin shank but a large, flat head to prevent punching through the drywall (Sheetrock) paper surface. The shank is usually deformed.

Annular ring and *screw shank* nails are used mostly for nailing underlayment over a plywood subfloor. For this reason they are sometimes called underlayment nails. They are also extremely useful in nailing down finish floors (screw shank) and in nailing any plywood surface to green framing. Illustration 65 shows how

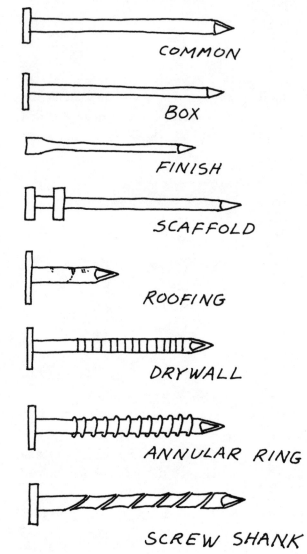

COMMON

BOX

FINISH

SCAFFOLD

ROOFING

DRYWALL

ANNULAR RING

SCREW SHANK 64

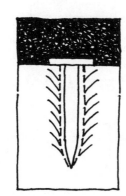

65

for exterior nailing to prevent staining of siding — even if you plan to paint or stain later.

It has been found that hot-dipped galvanized nails surprisingly have no significantly greater withdrawal resistance than bright nails. Apparently the rough zinc coating tears the wood fibers enough to compensate for any potential gain.

the annular ring nail retains its holding ability as green wood dries. When wood shrinks, strange as it may seem, holes in the wood get larger. They never get larger than the diameter of the annular rings, however. Common nails may lose two-thirds of their holding power as wood dries, while ring nails lose virtually none.

Cement-coated nails, sometimes called "sinkers," are box nails dipped in a resin. When the nail is pounded in, the resin melts. Upon cooling, the resin acts as a glue, increasing holding power by as much as 100 percent. However, the resin loses its holding power over time, and particularly after once letting go. Don't ever use sinkers to fasten a finish floor, because once the nails yield they become extremely noisy. They are ideal, however, for applying temporary glue pressure to gussets, box beams, etc.

Nailing Geometry and Load Holding

Figure 66 shows four ways of nailing two pieces of wood together. Also shown are the relative resistances of the nails both to lateral force and withdrawal force. The arrows define the directions of lateral and withdrawal force.

Sheet Metal Framing Aids

A number of prefabricated galvanized steel devices are available that speed installation and often improve the strength of the connection. Most of these devices convert a toenailing operation into facenailing. The most common such device is the "joist hanger," essentially a stirrup that supports a joist from both sides and bottom, thereby increasing its resistance to splitting at the end. It is available only for dressed lumber sizes. To use it with rough lumber requires sanding or otherwise reducing the

Bright vs. Galvanized

A plain steel nail is known as "bright." And it is until it rusts! Hot-dipped galvanized nails are dipped in zinc and will not rust. Other nails that will not rust are electroplated, aluminum, and stainless steel. *Always* use rustproof nails

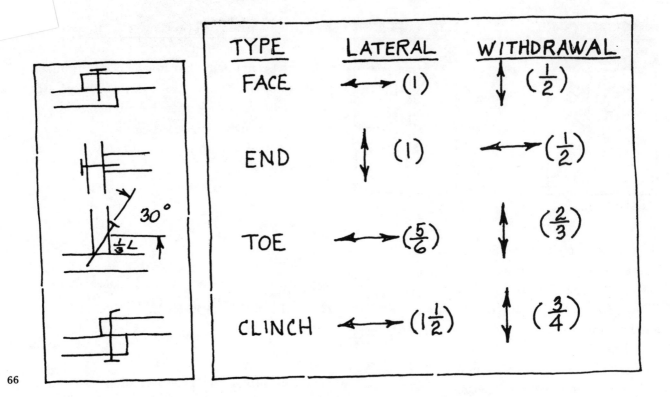

TYPE	LATERAL	WITHDRAWAL
FACE	←→ (1)	$\updownarrow$ ($\frac{1}{2}$)
END	$\updownarrow$ (1)	←→ ($\frac{1}{2}$)
TOE	←→ ($\frac{5}{6}$)	$\updownarrow$ ($\frac{2}{3}$)
CLINCH	←→ ($1\frac{1}{2}$)	$\updownarrow$ ($\frac{3}{4}$)

66

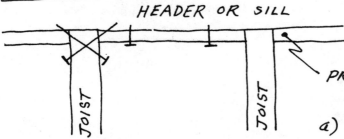

HEADER OR SILL

JOIST   JOIST

67

PRECUT 14" LONG 1" ROUGH
BOARD BLOCKS SERVE TO:

a) SPEED OPERATION MAKING
TOENAILING A CINCH

b) DECREASE SPLITTING OF JOIST

c) PERMIT TOENAILING FURTHER
BACK WITH LARGER NAIL,
THEREBY INCREASING
JOINT STRENGTH

Table 8. *Framing and Sheathing Nailing Schedule*

| Component | Method | Nails for | |
		S4S Lumber	Rough Lumber
Header to joist	end nail	3/16d	3/20d
Joist butted to sill	toenail	3/10d	3/12d
Joist cantilevered over sill	toenail	2/10d	2/12d
Bridging	toenail	2/8d	2/8d
2″ T & G floor to joist	facenail	2/16d	2/20d
Soleplate to stud, on deck	end nail	2/16d	2/20d
Soleplate to stud, vertical	toenail	4/8d	4/10d
Top plate to stud	end nail	2/16d	2/20d
Soleplate to joist or header	facenail	1/16d	1/20d
Built-up beam, 3 members	facenail	2/30d @32″	2/60d @32″
Let-in brace, 1″	facenail	2/8d each stud	2/10d
Collar beam to rafter 1″	facenail	3/8d	3/10d
Rafter to ridge board	toenail	4/8d	4/10d
Rafter 2″ to top plate	toenail	3/8d	3/10d
Rafter 4″ to top plate	toenail	3/12d	3/16d
Rafter to ceiling joist	facenail	4/10d	4/16d
Plywood subfloor 1/2″	facenail @6″	6d annular ring	6d annular ring
subfloor 5/8″	facenail @6″	8d annular ring	8d annular ring
wall sheathing 3/8″	facenail @3″	8d annular ring	8d annular ring
wall sheathing 1/2″	facenail @3″	8d	8d
roof sheathing 3/8″	facenail @6″	6d	6d
roof sheathing 1/2″	facenail @6″	8d	8d
Roofers, T & G	facenail	2/8d	2/8d

Note: All nails are to be galvanized common except as noted. If much splitting occurs, use galvanized box nails or blunt the ends of common nails.

2-inch thickness to about 1-1/2 inches. For this reason, plus the added cost, I sometimes use the trick shown me by an old-timer (Illustration 67).

One place where the extra cost and sanding time are worthwhile is in an exterior deck or anywhere else where two pieces of wood meet to form a potential water trap. If the construction of Illustration 67 were exposed to the elements, rainwater would seep between wood surfaces and soon rot the wood. A galvanized joist hanger permits standing the joist 1/4 inch back from the header, permitting air circulation.

Check with your local lumberyard to see if any other framing devices are available that might either save time or prevent a bad case of wood rot. If they are too expensive, consider the local sheet metal shop. They generate large amounts of scrap that they will either give you or sell per pound.

(CW)

The Building

14

Foundations

A close friend of the family gave me my first job as a carpenter. Only a close friend, and a courageous person, would have done that. I had no experience I could point to, only my word that I knew enough to get the job done.

The job Mrs. Morris gave me was the installation of French doors in place of a small window. I thought, measured, ordered the doors, understood what needed to be done, and asked some questions (not of Mrs. Morris) about how it should be done.

Then I went home, got a sledgehammer, came back to the good woman's house, and swung the sledge mightily, knocking a large hole in the wall below the window that had to come out.

"Oh!" exclaimed Mrs. Morris, blanching at the havoc, "I didn't know that's how it should be done."

I didn't know either. But I was certain after I swung the hammer and the hole appeared that I would have to continue until I finished the job, or it finished me.

It's the same with foundations.

Once the shovel bites the ground, once the concrete is poured, the posts put in place or the stones piled, your shelter becomes three-dimensional. The job has begun. It has got to continue until the walls go up, the roof goes on, and the furniture is moved in.

Until the foundation, all is theory, or one-dimensional. Yes, you have the land. But if you have not built, chances are you could sell it for more than you paid. You have done the planning and design work, but you can roll blueprints and put them on a shelf. The mortgage can be canceled, as can the lumber order, the nail shipment, and the advertisement that your current home is for sale (or the notice to your landlord that your lease will not be renewed).

None of these is small potatoes, but they are definitely midgets compared to the actual layout and construction of a foundation.

Trees crash to the ground, the integrity of the soil is violated, a man-made structure appears in a place that may have been in its essential natural state for hundreds, perhaps thousands, of years. And you are responsible.

The foundation is the beginning of your shelter. It is an irrevocable commitment, it outlines on the earth a presence that had always before been outlined on paper. It allows you to show people the place and say, "This is the house. We are standing in the kitchen." And when the first post or wall is constructed, it gives you a place to jump up on; it puts you at the height your living room floor will occupy, lets you peer from imaginary windows and close imaginary doors. The start of foundation construction is akin to taking the oath of office, hoisting anchor, turning the key, casting the fly, and saying "I do." It is the move that begins a series of scores of other moves, none of which can be stopped, all of which stretch before you in such profusion that they bend over horizons too distant to be perceived.

A foundation also holds up your house, and therefore is of certain importance in its own right.

I have never fully understood how a foundation holds up a house. In my state, Maine, I see huge barns built two centuries ago standing firm on tenuous pillars of granite rocks piled one atop the other by some farmer in 1770. There is no mortar anchoring the rocks, there is no visible bond between the barn timbers at the topmost rock. The barn merely "sets" there, looking as if a good wind could slide it off its precarious perch. Yet many good winds have blown in the century just past, and although the barn no longer serves a vital function, it stands tall, with or without foundation care.

I have seen, as you must have, buildings like these barns everywhere in the nation. There is only one uniform standard that binds them: they are old buildings, put up before the lunge toward technology that led to poured concrete and full cellars.

Some basic skill must have been shared by those long-ago builders, and it is the precise components of that skill that I do not fully understand. I do, however, believe I can identify one of the universal premises shared by those foundation masters of the past.

Because they had not arrived at the notion that technology could "master" nature, they designed and planned foundations that worked with the natural forces, rather than in defiance of them.

That is still the essential premise to keep in mind when you plan your foundation.

Consider your particular soil, your climate, your area's potential for frost, rainfall, squalls, hurricanes, snowstorms, and other verities of the annual cycle. Recall that wood, that most wondrous of natural materials, needs to be kept ventilated and dry. And consider the astronomical cost of hiring a bulldozer to excavate, the trauma such an excavation brings to the ecology of your site, and the ugliness you will have to endure as you live with the huge mound of material the bulldozer piles up as it excavates deeper and deeper for your full cellar.

It is difficult for me to comprehend how any harmonious relationship with the natural world of your site could lead you toward a full cellar. One does not follow the other.

That leaves you with a choice of a concrete slab, granite blocks, or posts of one kind or another, each of which, by the way, is given short shrift in the next chapter.

There is clay and bedrock and a high water table where our home stands. There is also frost of fair intensity. Our home stands on fifty-eight concrete posts poured onto a large base, rather like an upside-down mushroom. The posts hold up huge beams that hold up the floor joists that, in turn, provide a base for the floor. The floor line ended up being a minimum of 2-1/2 feet above the ground, and an ingenious sort of "skirt" around the house — designed to be invisible as such from the exterior and interior — closes in a crawl space under the house. The ground there does not freeze, thus the frost cannot push. The groundwater is not molested by submerged walls, however, so it flows under the soil under the house in quiet and harmless fashion. Small ventilating doors at each end of the house are opened in the summer to allow the prevailing winds to keep the timbers dry and ventilated; in the winter the doors are closed, and forced hot air from the furnace takes over.

Because we wanted to work the least damage to the area's natural integrity, we were able to save trees growing within a foot of the house. That is almost impossible with a bulldozer, and quite impossible for most bulldozer operators. The timbers that sit on the posts that hold up our house had already held up one of those barns for a hundred years. It has been three years and a score of storms and freezes since our place was built. Like the barns, it hasn't moved an inch, no pipes have frozen, and no rot has become apparent.

None of which keeps me from worrying. Nothing has been devised that can do that. I am satisfied, however, that we have the best foundation for our particular circumstance. You can have the same if you really push for the truth about the natural realities of your site. Once you understand them, your foundation will design itself, which is the point of the following.

Until your foundation is begun, until that first shovel turns, you may never comprehend the significance of the moment. It is much more shattering than a sledgehammer swung at the wall.

By the way, I got out of that all right. The French doors worked. They are still working, as far as I know.

(JNC)

15

Foundations—
Some Details

No other single aspect of house building seems to evoke such intense emotion or fuzzy thinking as the foundation. Nine people out of ten consider a house without a full concrete basement as low-class construction suitable only for vacation homes or farm sheds. The one remaining person is open-minded enough to consider the concrete slab. I am secure enough in my thinking to ignore the opinion of the general public; but when this opinion extends to local building codes and mortgage officers my blood begins to boil. Building codes generally have two stated functions: to ensure safe construction and to prevent construction destroying the general integrity of the community. The National Building Code, based upon safe engineering practice, is fairly free of prejudice and stops short of specifying the types of foundation a house must have. Local building codes, however, are all too often based entirely upon hearsay evidence, the prejudice of a few individuals who proposed them, or the wish to prevent the spread of low-cost (read low-class) housing.

The primary responsibility of the mortgage officer is to the bank and not, as the advertising claims, to you. That responsibility reduces to one consideration — resale value. Resale value, of course, is what the public is willing to pay. Thus, we have come full circle. I don't say the banks must *promote* alternative foundations, nor do they have to assign equal resale value to houses regardless of foundation, but their present attitude (no mortgage, period) is clearly the stopper in the pipe. Anytime a convention of bankers wants a speaker, I'm available.

The Origin
of the Full Basement

How did we arrive at this point — where a house by definition, sets over a full basement? Rex Roberts's explanation is so good I have to borrow from it. Everyone knows that wood, lying on the ground, is attacked by soil fungus in a few years. Fence posts set in the ground, even cedar ones, rot and break off at ground level within about ten years. Therefore the wooden sills (bottommost beams of a house) are better off if set upon large stones. In the old days they didn't have the insulating materials of today or even building paper that would last any length of time. Therefore, the wind under the house was accustomed to rushing up between the floor boards. If the stones holding up the sills formed a solid wall, the wind velocity was at least cut down.

Great-grandfather used to produce about 70 percent of all his food needs right on the farm. Vegetables (particularly potatoes) were best stored at temperatures just above freezing in high humidity and in total darkness. The root cellar — essentially a well in which water never accumulated — served this purpose perfectly. The most convenient location for the root cellar was right under the kitchen, reached by a trapdoor. After a while the practice of canning allowed the preservation of an increased variety of foods — even meats. More food to store required a bigger root cellar (by now known simply as the cellar).

Heating the house with wood used to be a dreadful chore. The typical farmhouse, being poorly insulated and as drafty as a barn, had three or four wood-eating stoves that required stoking about every four hours. The average wood consumption for a year was around 15 cord! The advent of the furnace (wood or coal) allowed heating of the entire house from one central burner. The only way of moving heat at that time was by the natural rising of hot air. And the only place from which hot air could rise was the cellar. The cellar also had plenty of space for the ductwork. Unfortunately, the furnace heated up the cellar, too,

which ruined the potatoes, but the house *was* warmer. Fifteen cord of wood is a lot of wood, and it couldn't be stored in the cellar because the cellar was damp. But coal required less volume and it could be stored in a wet location. You can see that the cellar had, by this time, a true function. Sometime later the oil furnace arrived. It replaced the coal furnace, so it, too, went in the cellar. And the gas furnace too. (Fortunately, furnaces today are so clean and safe that they could be built into a closet.) When the automatic washer and dryer arrived, no one knew quite what to do with them. They were noisy, leaky, and required new plumbing and wiring. Obviously, the cellar was the perfect place.

With the arrival of the oil furnace and the supermarket, our fathers were left with an enormous space containing only an oil burner, a small oil tank, and a washer and dryer. It wasn't long before the weekend handyman, Ping-Pong players, and beer-drinking TV-watchers were hustled down into the out-of-sight hole-in-the-ground. The handyman and beautiful house magazines have had a running contest for decades now over how to make a cellar seem not a cellar. The latest one I read had a family room with a *fake skylight* lit by fluorescent bulbs!

There are no two ways about it — I agree with Rex Roberts — the cellar is second-class space. It's naturally dark and damp, and to achieve anywhere near the same livability as above-ground space, it will end up costing as much or more per square foot.

The Function of the Foundation

I hope I have made one thing perfectly clear: a basement is not a foundation and a foundation is not a basement ... necessarily. Then what is a foundation? As with most things in life, the true definition is best achieved through examining its function. Architects and architectural critics have gotten all bogged down on whether form is function or whether form follows function. One great man makes a perfectly clear statement and all of the hangers-on fight over it like so many chickens with a worm.

We won't get involved in the pedantics but will get straight to the heart of the matter. The form our foundation will take should be dictated by the functions we expect it to perform:

To transfer and distribute building loads to the ground.

To prevent building uplift, overturning, and horizontal sliding.

To keep the building high and dry.

To prevent motion due to frost.

TO TRANSFER AND DISTRIBUTE LOADS

A building weighs a great deal. We have already discussed building dead loads, live loads, and snow loads. It is not unlikely that the average house at some time in its tenure will impose a load on the ground of several hundred thousand pounds. The ability of the ground to support this load without yielding is called the soil-bearing capacity. To get a feel for the loads the ground is subjected to, calculate the load under your heel each time you take a step. My heel measures 3-1/8″ × 3-1/4″, or about 10 sq.in. I weigh around 140 lb. Therefore, the soil pressure under my heel is:

$$pressure = weight/area$$
$$= 140 \; lb./10 \; sq.in.$$
$$= 14 \; psi, \; or \; 2,016 \; psf$$

When I step in the freshly plowed garden I sink in a few inches. At the beach, in the tidal zone, I seem not to sink at all. In other words, different soils have different bearing capacities. The National Building Code specifies that, in the absence of actual tests of bearing capacity, we should use a *presumptive soil-bearing capacity,* as follows:

soft clay	3,000 psf
medium clay	5,000 psf
fine sand	4,000 psf
coarse sand	6,000 psf
gravel	8,000 psf
bedrock	30,000 psf

The only trouble with the presumptive table is that it presumes an ability on our part to discriminate between mud, soft clay, and medium clay. Nice, firm clay that a pickax bounces off in the summer can be bottled in the spring after you muck around in it a little. What to do other than search for bedrock? Easy — increase the size of our feet! That is, provide the foundation with a large enough *footing* to hold the soil load to a small number. I recommend:

(1) Don't build at all in organic-type soils, such as peat, or in filled-in areas.

(2) Build single-story houses with soil loading of 2,000 psf maximum in all clay soils.

(3) Restrict yourself to 3,000 psf in all other firm, undisturbed soils.

(4) On bedrock, don't worry.

In Chapter 9 we learned how to calculate the total deadweight of a building. If we simply add the snow load and live loads we have the design maximum load on the footings. Large buildings are often constructed with the expectation that they will settle, oftentimes several inches. There is nothing wrong with a small degree of settling per se — it's only uneven settling that gets us in trouble, à la the Tower of Pisa. Therefore, size the footings so they all carry the same *design load pressure* (psf). Chances are, the building will settle a fraction of an inch, but you'll never notice it.

TO PREVENT BUILDING UPLIFT, OVERTURNING, AND HORIZONTAL SLIDING

We have already learned how to analyze a building and its foundation for stability in high winds. Remember, however, that for a foundation to act as an effective anchor: the building must be anchored to the foundation; the foundation may have to be anchored to its footing; and the weight of soil directly above the footing may have to be included.

If building directly on bedrock with no overburden, we have two choices: drill into the bedrock and pin the foundation to the rock using an expansion bolt, or provide all the resistance to overturning with the sheer mass of the foundation. Personally, I feel a lot better about the latter option. I don't trust what I can't see (the bolt) and concrete is a bargain at 1/2 cent per pound, delivered.

TO KEEP THE BUILDING HIGH AND DRY

For keeping a building high and dry, concrete is the next worst thing to setting your sills directly on the ground. Concrete attracts water and will actually lift water from the ground to keep the sills damp. Either treat sills with a penta-type preservative or place a waterproof barrier such as polyethylene between sills and concrete. In the case of a low-to-the-ground slab foundation, use sills that have been commercially pressure treated. Otherwise keep the sills a minimum of 12 inches off the ground. This will help keep them dry from roof drip, splash, snow, and that nice dewy grass you are always meaning to clip.

TO PREVENT MOTION DUE TO FROST

You have probably seen milk frozen in a bottle or a pipe burst by freezing. Water expands upon freezing a great deal — about 4 percent by volume, to be exact. To make matters worse, when water freezes in the ground there is nowhere to expand but up, since down and sideways are already occupied (except in the case of full basements!). This means that the surface of ground that is saturated with water actually moves up 1/2 inch for every foot of frost penetration. Clay is notorious in this connection because of its attraction for water. Water can be pulled as much as 30 feet above the water table by clay. Clay is therefore almost always saturated. Silt is even worse, forming ice lenses. On the other hand, if there is little moisture in the ground — that is, the ground is well drained — the minimal ice has plenty of interstitial space in which to expand and the ground doesn't move. Therefore, well-drained sands and gravels are not subject to frost heaving.

Illustration 69 is a pictograph showing the mechanism of frost penetration. We make several simplifying assumptions but the basic

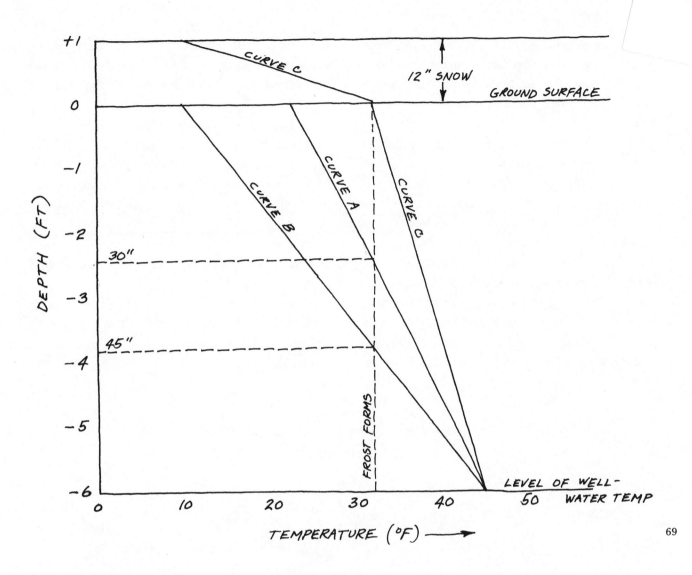

physics is correct. I show it to you so that you will know how to think about frost.

The top of the pictograph represents the surface of the ground. The temperature there varies from about -20° F to 100° F, depending upon the weather. Not too far down on the chart (-6 feet) the temperature of the ground varies only a few degrees from well-water temperature, which is approximately the average annual temperature in the locality plus 2 F°. On the graph I have shown that temperature for my site, 45° F. Freezing occurs when the ground temperature reaches 32° F. If we make the simplifying assumption that the

ground is homogeneous, and that its thermal conductivity is therefore constant with depth, then the temperature in the ground will be found on the straight line connecting the average surface temperature and the constant well-water temperature. This assumes further that the average surface temperature holds for several weeks. For example, with bare ground and an average January temperature of 23° F, the temperature line is shown as line A. The depth of frost is found as the depth at which line A crosses 32° F, roughly 30 inches.

Note that, provided the average temperature goes no lower, the frost will penetrate no further. Suppose that a prolonged period of intense cold occurs with an average temperature of only 10° F. Then the ground temperature slowly moves toward line B. The depth of frost for line B is 45 inches.

The farmers have a saying to the effect that an early snow which stays late means no frost. How can that be? The thermal conductivity of dry snow is less than one-tenth that of the ground. Recall in fact that 11 inches of new powdery snow has the same insulating value as 3-1/2 inches of fiberglass insulation. The better the insulator, the greater the temperature drop within the material. The temperature inside the snow will change ten times faster than in the ground, leading to line C. Of course the ground surface must ultimately become 32° F or the snow will melt from the bottom up. Even so, the portion of temperature line C belowground is everywhere above 32° F, so the farmers are right.

We can make other observations from Illustration 69. Water pipes passing under driveways should be buried several feet deeper because: 1) the snow cover is removed, and 2) the soil is compacted, making it a poorer insulator. Farmers get angry at snowmobilers because the compacted snow of their tracks ruins the insulative value of snow and causes frost to remain longer in the ground.

Of course our main concern with frost is regarding the foundation. What can frost do to a foundation? At the very least it can push the foundation up if it penetrates beneath the

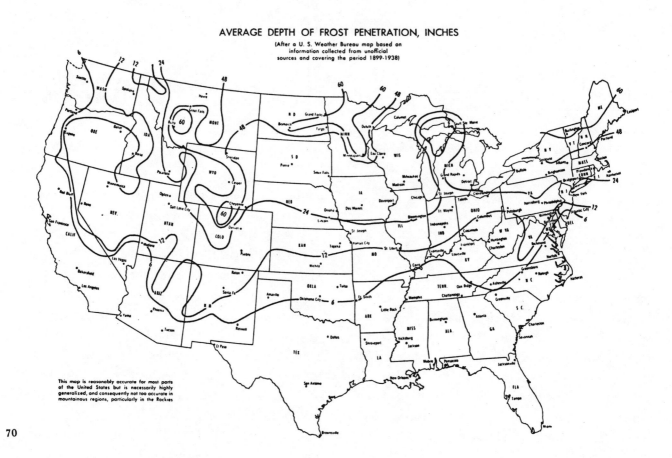

AVERAGE DEPTH OF FROST PENETRATION, INCHES

(After a U. S. Weather Bureau map based on information collected from unofficial sources and covering the period 1899-1938)

This map is reasonably accurate for most parts of the United States but is necessarily highly generalized, and consequently not too accurate in mountainous regions, particularly in the Rockies

70

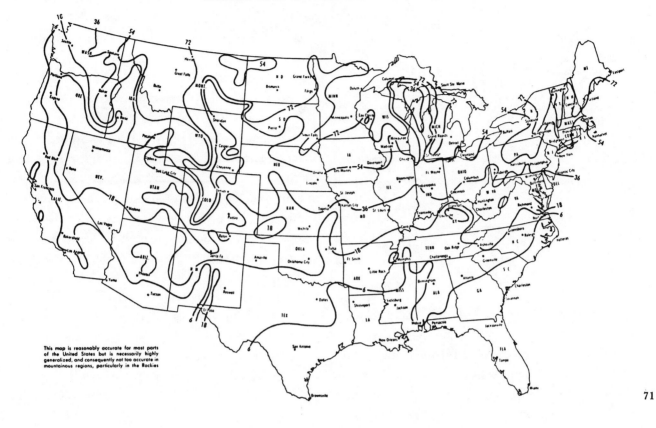

This map is reasonably accurate for most parts of the United States but is necessarily highly generalized, and consequently not too accurate in mountainous regions, particularly in the Rockies

71

footing. Illustration 70 is a map of average depth of frost penetration. If you live in an average area, use it as a guide. If you live on a windswept hill where the snow never accumulates, use Illustration 71.

There are three more considerations you should be aware of. If you build absolutely without foundation in a poorly drained soil (the National Building Code allows this for buildings of less than 500 sq.ft.), you are likely to make vertical excursions of more than an inch each year. But, worse still, when frost leaves the ground it leaves empty spaces and enough water to make a mush that has about enough bearing capacity to support a Ping-Pong ball. You will find your house slowly sinking into the ground a little more each year.

The second consideration is the particular combination of frost, clay, and concrete "sono-tubes." When the frost expands it grips the sides of the concrete post. As the ground surface rises so does the post — unless either the post is anchored to a large footing that will hold it down or the post tapers upward. The third problem also involves concrete posts in clay and is shown in Illustration 72. Due to the warmth of the sun, the ground on the north side of a post generally freezes a bit earlier and deeper than on the south side. Each year the post is pushed a bit more from the north than from the south. Clay is a plastic material that yields to pressure much more readily than other soils. Suppose the post moves 1/4 inch per year. After ten years it will have moved 2-1/2 inches. Any tilt from the vertical results in a *moment* consisting of the weight on the post times the

FOUNDATIONS — SOME DETAILS / 125

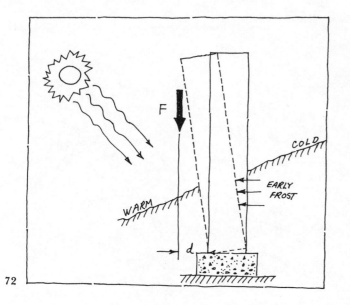

72

is the prevention of backsplash against sill and siding, which can lead to rot and stain on natural or stained siding. Sloping the ground away from the house in all directions disposes of the rain falling on the ground near the house. This is particularly effective with a clay subsoil since clay is impervious to standing water. Lastly, and most important, the absolute frost-free guarantee is provided by drain tiles that collect and eliminate water, whatever its source, from the area of the footing. The drain tiles must slope continuously toward the discharge end; otherwise sediment may build up at a low point and completely block them. This method is impossible on a perfectly level site. If your site is flat and level and you aren't convinced yet of the advantages of the southern slope, the best you can do is to dig deeper, anchor to a big footing, use tapered posts, and backfill with the originally excavated material. If you are in clay, filling the hole with gravel will only result in a gravel-filled well.

horizontal displacement of the top. After a while this moment itself is enough to cause a steady plastic migration through the clay, resulting in ultimate disaster. Wooden posts are easier to brace against this moment, and for this reason are superior in clay. They also have an elasticity that allows them to snap back after the thaw. Sometimes an extra-large (or even continuous) footing is used with concrete piers, each pier having a "moment connection" by way of a reinforcing rod to the footing.

Drainage

Remember that frost can only give us trouble if the ground is nearly saturated with water. Drainage is the prevention of this condition. Illustration 73 shows the elements of proper drainage. Roof gutters, downspouts, and leaders can discharge all of the water falling on the roof far from the building. An additional benefit

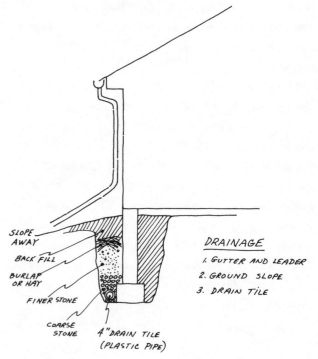

DRAINAGE
1. GUTTER AND LEADER
2. GROUND SLOPE
3. DRAIN TILE

73

Frost is not the only reason to be concerned with drainage. A full basement foundation is essentially a large and expensive concrete-lined well that we try to keep dry. This proves to be surprisingly difficult, and I will show you why. Illustration 74 shows a full basement under a 20' × 32' house. Every spring, just after the snow melts, the ground water table is at its highest. (Wherever you see alder trees growing you can be sure that the seasonal high water table is identical with the ground surface. Don't consider for a moment building where alders thrive!) Suppose that we neglected the drain tile. Then our basement resembles a concrete boat immersed in a mixture of soil and water up to the water table. Water exerts a pressure in all directions equal to the weight of water in the column above. Water weighs 62.4 lb./cu.ft., and therefore, the water pressure at depth is: $p = 62.4$ *psf* × *depth in feet beneath water table.*

In Illustration 74 the average depth under the floor slab is 6 feet, so the pressure of the water trying to get into the basement is 375 psf, or 2.6 psi! No wonder they call basement waterproofing paints "miracle paints"!

The vertical lifting force (or buoyancy) on the basement is 375 psf × 20' × 32' = 240,000 lb.! This is greater than the weight of the average house. The MIT area of Cambridge is filled land reclaimed from the Charles River. The water table here is essentially constant — that of the river. Multistory concrete dormitories have been built there using the buoyancy of the basement to float the entire building.

In residential construction a drain tile is placed around the perimeter of the footing. The water table is held at the level of the tile under the entire basement so the water pressure is held to zero. Study the installation of the tile carefully. Note that, once again, a naturally sloping site allows natural drainage. On a level

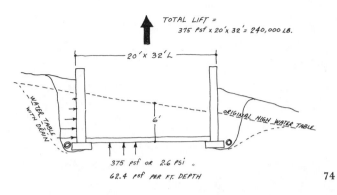

74

site the drain tile would have to be discharged into the city sewage line, or a drywell and sump pump installed.

Foundation Options

A number of foundation options are open to you. I will discuss six, which cover the spectrum. Which one you choose depends upon your budget, your feelings regarding concrete and wood, your opinion of belowground living space, and perhaps, unfortunately, your local building code.

FULL BASEMENT

The full basement is the option with which, due to familiarity, most of us are comfortable. Technically, however, it is among the most questionable. As I showed earlier, the basement foundation is essentially a concrete well or boat, which we hope will remain dry. Even if our drainage precautions work, the basement is still likely to be a rather damp place due to sweating of the concrete walls and incoming water pipes.

A lot of people hold the opinion that the full basement is the only way to achieve a warm first floor. At the time this opinion originated it was true. Before the advent of modern insulating materials, the insulation of the earth itself was the best available. The thermal resistance of earth on a per-inch-of-thickness basis is extremely poor. However, in a full basement we are effectively surrounded by a great number of inches of this material. The thickness is not the distance to China, however, but only the distance to the well water temperature, or about 6 feet. The rule of thumb used by heating engineers is that the full basement has an effective thermal resistance of R10. (R20 would be twice as great and result in half the heat loss, annually.) Now R20 can be achieved by a wood floor and either 4 inches of plastic foam or 6 inches of fiberglass insulation. Furthermore, unless we provide heat specifically, insulating the floor over a full basement will result in a cold basement, which voids the living-space reason for the basement.

Still other people look to the basement as an inexpensive storage area. To them I recommend Thoreau on possessions: "We are prisoners of our possessions. . . ." Beyond the philosophical argument, unheated storage space can be built above ground at half the cost.

I can think of only one valid reason for the basement foundation — the root cellar. There is a revival of interest in the preservation of home-grown food both raw and canned. Of the items commonly stored, however, only the root vegetables and apples store well in the cool and damp environment. Squash and onions are better stored at moderate temperature and low humidity. All the vegetables and canned goods for a year for a family of four could be stored in a root cellar measuring 8 feet square and it doesn't have to be under the house!

PERIMETER WALL

The perimeter foundation wall makes a bit more sense because it's doing one thing adequately (holding up the house) and not several things poorly. It also represents a step toward more rational design in using *less* material for the specified function. However, if the design criterion used is cost versus effectiveness of function, the options described below have it beat. Unfortunately, it is the minimum specified by all too many local building codes.

SLAB ON THE GROUND

The concrete slab is very popular because it promises to fulfill at reasonable cost two major requirements: foundation and first floor. As a foundation it is outstanding because, acting as one gigantic footing, it imposes the lowest soil loading in psf of all foundations. At the same time, its enormous weight is insurance against wind uplift.

As a floor, the slab has had its ups and downs. The earliest slabs got a bad reputation for being damp. This problem can· be overcome by proper site preparation and the use of a polyethylene vapor barrier. Never install a slab without at least a continuous 6-mil polyethylene membrane — and be careful that it is not punctured during the pouring of the slab. Water can still slip between the membrane and the slab at the perimeter. Thus, it is important that the ground slope away from the slab in all directions at the minimum slope of 1 foot in 25. If the slab is to be poured over a clay subsoil, 2 feet of clay should be excavated and replaced by a layer of gravel and sand of uniform thickness compacted by a vibrating compacting machine in 3-to-4 inch layers. This will stop the moisture rising in the clay and ensure against frost heaving if the building is ever left unheated.

Most of the heat loss through a slab occurs at the perimeter. For this reason, rigid insulation should be placed around the perimeter. Heat loss through the bottom of the slab will still be proportional to the R10 effective resistance value of the earth unless further insulative measures are taken. One possibility is pouring the slab over rigid foam insulation panels or insulation foamed in place on the ground. The other possibility is laying a plywood floor over foam that has been laid on the concrete using mastic cement.

The slab disappointed others who thought they could neatly dispose of all that ugly plumbing and wiring by pouring concrete over it. Unfortunately, pouring concrete over anything doesn't leave much flexibility — sorry about the pun. They found that when the pipes burst due to differential expansion and contraction and just plain corrosion, the appropriate tool was the jackhammer instead of the pipe wrench.

Actually, the slab can be one of the most versatile and practical foundations of all. Here's how I would use it. First I would take all of the precautions necessary to ensure a dry, frost-free slab (ground sloping away, 2-foot compacted gravel bed, polyethylene membrane, and perimeter insulation). Then I would build a floor platform 16 inches above the concrete, joists and all. The joists can be as small as 2″ × 4″ because blocks are easily provided from the concrete below to limit the clear spans. Plumbing and wiring can be run under this platform in any direction, just as in a full basement. With trapdoors, this space provides an incredibly large storage space. Being isolated from the heated living space, it is a cool, dry space not too bad for storing food. If you line the top and bottom of the space with reflective aluminum foil, the insulation value of the floor/slab combination goes up to about R20. On the other hand, if you use forced hot-air heat, the space can serve as one gigantic hot-air duct. To get heat you need only open a small louvered floor register.

The entire floor need not be raised 16 inches. A sunken floor effect can be achieved by raising some area less. It's even fun to contemplate someday redesigning and remodeling the entire floor. It would be simple since the exterior walls don't rest on the floor.

BALLAST BED AND RUBBLE TRENCH

Have you ever wondered why railroad tracks don't get the frost heaves, or why some highways have none while others get terminal cases every year? Good highways and railroad beds are built in the same way. First the roadway is graded for proper drainage. That is, the roadbed is built up and sloped so that surface water immediately drains away. Drainage ditches along the edges handle the maximum surface runoff. At low points, culverts are placed to eliminate excessive ponding. These steps guarantee that the road foundation will not get mushy and provide an adequate soil-bearing capacity. Next, a layer of crushed stone (ballast) is spread. Finally, the road is compacted and surfaced or the railroad ties are tamped firmly into place on the crushed stone. The function of the stone is threefold in the case of the railbed:

Being extremely coarse it never retains water.

It distributes the concentrated load on the railroad ties not straight down, but at a 30° angle from vertical to a wider area of the subsoil.

It guarantees by its thickness that frost will never reach the subsoil to cause heaves.

Frank Lloyd Wright in *The Natural House* describes a similar method used by Welsh stonemasons. He recommended placing large rubble (fist-sized broken stone) in trenches 16 inches deep on which the foundation blocks or stone walls are then placed. Wright's recommendation of 16 inches assumes either a frost depth no greater than 16 inches or a well-drained subsoil. I will show two modifications of this method that should work well regardless of frost depth or subsoil type.

On a sloping site excavate an area about 1 or 2 feet larger than the house all around. The depth of excavation depends upon frost depth and how much stone you want showing above the natural ground. Place a layer of crushed or screened stone of thickness equal to the maximum frost depth. Make sure that the excavation is level or slopes slightly downhill. If the excavation doesn't break the surface downhill but resembles a shallow swimming pool, place drain tiles in the bottom of the pit that drain to the surface at some point downhill.

On top of the crushed stone place, or cast in place, large concrete blocks the masses of which are sufficient to prevent building overturning in the wind. Anchor bolts are cast into the blocks for attaching the sills.

If crushed stone is too expensive, the amount required can be reduced by excavating 2-foot-wide trenches under each sill, which are then filled with the stone. Once again, don't forget the drain tile if needed.

The concrete blocks are easily cast using plywood forms. The blocks can be cast in place to any height for a sloping site. If the desired height is over 24 inches, install form ties or wrap the form tightly with steel wire to prevent bursting. As usual, tamp the concrete well as it is poured to prevent ugly air pockets.

Illustration 75 shows a 24″ × 24″ × 16″ block and form.

The resulting block has a handsome molded appearance because of the corner strips of wood. Even fancier blocks can be made by nailing other wood strips inside the form. Three blocks can be made using a single piece of 1/2-inch CDX plywood and the forms can be reused. Each block weighs 800 pounds. The concrete for each costs about $6 delivered at $30 per cubic yard.

I like this foundation because: 1) it's as sure as any to solve the frost problem; 2) it's cheap; 3) it's noncommittal — the blocks are easily moved, replaced, added to, or shimmed if for any reason they should require it; and 4) it

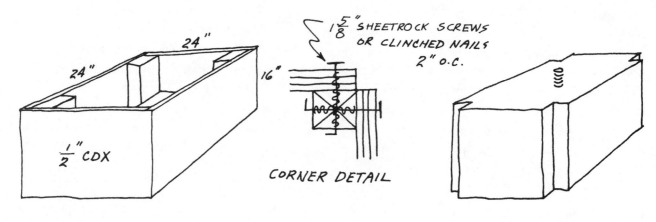

75   FORM   CORNER DETAIL   FINISHED BLOCK

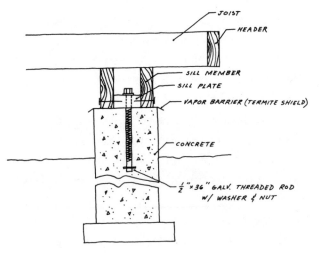

JOIST
HEADER
SILL MEMBER
SILL PLATE
VAPOR BARRIER (TERMITE SHIELD)
CONCRETE
½"x36" GALV. THREADED ROD
W/ WASHER & NUT

76

makes its own footing. The crushed stone spreads the load so that the soil pressure is extremely low. If it can support a locomotive, it will surely support your house.

SONOTUBES

The concrete sonotube is a concrete-filled cardboard cylinder that is used in much the same way as telephone poles. The only reasons I can think of to use them instead of telephone poles are: 1) if you or your mortgage officer think telephone poles will rot in the ground, and 2) if you need the weight to hold your house down in the wind. In fact, concrete *can* deteriorate (spall) and the reinforcing bar inside rusts, so I'm not at all sure that the sonotube will last longer than the pressure-treated pole. Other than that, the telephone pole foundation is far simpler to install.

A question that arises every time either concrete-cast blocks or sonotubes are used is: "How do I fasten the wood sill to the concrete?"

Illustration 76 shows the best method I have found. First of all, the concrete members must be set in from the building outside lines (the floor joists are cantilevered over the sill) so that water never falls onto the concrete top surface. In addition, a vapor barrier (galvanized sheet metal, heavy or multiple layer of polyethylene, etc.) is placed between concrete surface and wood sill. If you live in termite country, the galvanized sheet metal can act as a termite shield at the same time.

The fastening trick is to cast 1/2" × 36" galvanized threaded rods with a nut and washer at each lower end at the time the concrete is poured. Before the concrete has set make sure all of the rods fall along the desired center line of the sill. Hold the rods temporarily in place by a wooden collar with a 1/2-inch hold that is nailed lightly to the form. After the concrete has set (about two days), remove the collars. Place the 2" × 4" or 2" × 6" sill plate shown in the figure over the bolts and strike with a hammer, thereby causing an impression of the bolt. Drill 1/2-inch holes at the impressions. Install the vapor barrier. Slip the sill plate over the bolts and tighten down. Final tightening should be left for about a week. The sill plate is now firmly attached to the concrete foundation. The on-edge sill members are now attached by facenailing to the sill plate using 30d galvanized nails every 24 inches.

If a splice is necessary in the sill plate, either cast two threaded rods at that point, or make it occur at a point where the on-edge sill members are continuous, thereby acting as the splice.

PRESSURE-TREATED POLES (TELEPHONE POLES)

The suggestion of sticking a wood pole in a hole in the ground brings a picture to mind of leaning broken fence posts, old rotting tree stumps, or, perhaps, a crumbling, soaking wet, crawling-with-bugs piece of wood found as a child. There have existed for about forty years, however, techniques for chemically treating wood to make it rotproof and insectproof. Telephone poles are now treated in gigantic pressure cookers filled with chemical preservatives so that the preservative penetrates every cubic inch of the wood. If one were to cut through such a pole, these usually oily preservatives would be found all the way to the center, sometimes to the extent of oozing out.

The only way such a pole will ever deteriorate is if the preservatives are leached out of the wood. Leaching occurs faster in air than underground and so a pole will sooner fail in air or right at the ground surface than deep in the ground. A conservative figure for the life of an 8-inch-diameter pressure-treated pole under the worst conditions is seventy-five years. If that doesn't satisfy your permanence criterion, I have good news for you. I have personally replaced a pole (it didn't fail — I just didn't like the location) in four hours, digging by hand. All of the perimeter poles under an average house could be replaced in one day by a crew of two and a backhoe.

When a power line is moved in widening a street or a pole is damaged in an encounter with a vehicle, the poles are rarely used again. The disposition of the poles is sometimes on a first come, first served, basis. A used pole rarely lies by the roadside for a full day in rural areas. In metropolitan areas the value of a pole is not generally recognized by either the population or the building code, which allows their support of skyscrapers (as pilings) but not single-family residences. Therefore, metropolitan dumps are excellent sources of used and damaged poles.

If you can't find enough sound used poles, new poles can be purchased from the manufacturer, whose name and address can be gotten by asking the local utility line department boss. Don't fall for timber advertised as "land-scaping timbers." These are usually produced by merely dipping the timbers in a tank of creosote. The depth of penetration rarely exceeds 1/4 inch and its life is very questionable.

An excellent blow-by-blow prescription for the installation of treated pole foundations is to be found in the government publication *Low-Cost Homes for Rural America,* an excellent self-contained manual of conventional low-cost house building. I will describe a similar approach to the pole foundation.

To lay out the foundation, you'll need a line level, a ball of strong string, a tape measure as long as the diagonal of your house, some 1″ × 4″ boards and 2″ × 3″ or 2″ × 4″ stakes (See

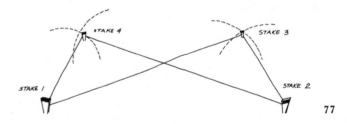

Illustration 77). After clearing (not bulldozing the site of obstructions, place small stakes 1 and 2 in the ground, representing the ends of the longest wall. Second, place stakes 3 and 4 at the intersections of strings of length equal to the side walls and diagonals. This will give you roughly the corners of the building. Now drive the large 2″ × 4″ stakes so that they box in the small stakes but 2 feet outside the building line (see Illustration 78). Using a level,

nail horizontal 1″ × 4″ batter boards at corner 1. The top of the batter board establishes the top of the sill.

Make sure that the height is such that the sill bottom will clear the ground by a minimum of 12 inches. Using the line level, nail the batter boards at corner 2 at the same elevation. Do the same to corners 3 and 4. Reverse the line level to detect any error in the level. We have now defined a level surface upon which the floor joists will rest. All that remains is finding the exact corners of the building in this plane.

Place nails at the midpoints of the batter boards in corner 1. Stretch strings from A and B to nails C and D in corners 2 and 4 (Illustration 79). Measure the length of the building

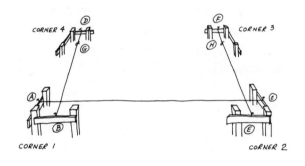

79

from point B and place nail E. Run a string from nail E to nail F. Measure from the intersection of strings in corner 1 the length of the side wall and mark with a piece of yarn point G. Measure from the intersection of strings in corner 2 again and tie yarn H. We now have the position of the front wall and the lengths (but not position) of the side walls. We now need the length of the building diagonal. This is easily arrived at by a number of methods: (1) Draw a large-scale triangle on a piece of paper and measure the diagonal using an architect's scale. (2) Use the Pythagorean Theorem (see Illustration 80). Use a calculator or slide rule or tables to find C, the diagonal. (3) Get a clever high school stu-

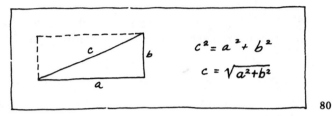

$$c^2 = a^2 + b^2$$
$$c = \sqrt{a^2 + b^2}$$

80

dent to do number 2 for you. I can't see why carpentry books are forever talking about 3, 4, 5 triangles. Anything but a doghouse is about ten times that large and the errors of extending the lines of a 3, 4, 5 triangle are precisely ten times the errors in measuring a 30, 40, 50 triangle directly.

Having the precise diagonal to an accuracy of 1/8 inch, have a friend hold the end of the tape measure at the string intersection in corner 1. Release the string from nail F and swing that string and the tape until yarn H falls directly on the diagonal measurement. Renail F to hold that string position. Repeat the process in corner 4 with nail D. Run a string so that it touches yarns F and H. We have now defined the four building corners by the four string intersections.

The accuracy of our layout can be verified by measuring the distance between yarns F and H, which should be the length of the back wall. If this is off by an inch or so, don't panic — it usually is. Remeasure and adjust all around until all the measurements agree to 1/8 inch, which is better than the housing industry standard. Besides, as you'll see in a moment, with a pole foundation it doesn't much matter.

If you agree with my way of thinking, your sills will be tucked in about a foot from the outside of the building, with the floor joists cantilevered over the sills. Run inside strings parallel to the original to establish the center line of the poles. Tie pieces of yarn at each pole location along the strings. When you start digging, you'll find the real value of the batter boards: The string is removed while digging, but can be reestablished anytime later without requiring remeasuring.

If the site has well-drained sandy soil, a hole for each pole makes the most sense. You can dig holes yourself or you can probably find a contractor specializing in pole buildings or sign erection who has a truck-mounted 24-inch diameter earth auger. If the soil is clay you should dig a trench for each row of poles with a drain tile in the bottom, which will later be filled with a coarse gravel. Pour a concrete footing for each pole below the frost depth and large enough in area to keep the soil pressure below the levels recommended earlier. If your over-turning-moment calculations show a need for including the weight of soil above the footing for stability, nail a row of galvanized spikes 6 inches from the bottom of each pole. After the pole is in place on the footing and lined up correctly, pour a 12-inch-thick concrete collar around the base of each pole. This will fix the location of the pole and make the soil above effectively part of the foundation.

In lining up the poles it will become obvious that telephone poles come in all sizes and shapes. Don't worry; since our joists cantilever over the sills, the sills are only required to be level, not straight. Snap a chalk line across the poles at the level of the strings. Then cut the poles off. Next cut lightly with a chain saw or buck saw into each pole, as in Illustration 81, to provide a seat for the sills. Nail a pair of sills to the poles. When you are satisfied with the location of the sills, drill clear through the sills and poles and bolt, using two 1/2-inch bolts

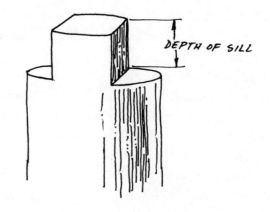

DEPTH OF SILL

81

per pole. If a sill splice occurs at a pole, use four bolts.

The sills provide the level surface upon which the floor joists are placed. The floor is framed simply lying on top of the sills. If using rough lumber for floor joists, you will probably find a variation of as much as 1/2 inch in the depth of the joists. Simply arrange the joists in order of increasing depth, and you will have a floor that slopes evenly by 1/2 inch across the length of the building — which is a lot better than 1/2 inch every 16 inches on center!

After all the joists have been connected by the headers, the floor platform is finally squared by "adjusting" the corners with a sledgehammer until the diagonals are equal. After the diagonals agree to within ±1/8 inch, nail the end joists to the sills. Using a string as a reference, straighten the headers with the sledgehammer and toenail each joist to the sills.

You're off and flying — aboveground at last!

(CW)

16

Framing

I had a concept of house building before we started our noble experiment. After the foundation, I rationalized, the essential house (walls, floor, and roof — minus plumbing, etc.) could be viewed as little more than a collection of lumber held together with nails. Scores upon scores of lengths of lumber, to be sure, and thousands upon thousands of nails; yet, if the building could be thought of as a series of boards, planks, sheets, or beams, cut to the right dimensions, and nailed to another board, plank, sheet, or beam, the entire process could be seen as the joining of each to the other.

Days broke down into hours, hours into minutes. In each recognizable interval, I could imagine a board being selected, sawed, notched perhaps, and then attached to another length of lumber that had already been put in place. If you think of the project in its smallest scale, I kept telling myself, you will not be overwhelmed by the enormity of your arrogance, the complications of the enterprise, nor the dimensions of the risks that ignorance entails. After

all, I continued (some days it took me as much as a half-hour of such pep talks before I could muster the gumption to visit the site and be greeted with a new crisis), you, in your dullest moments, can cut and nail a board.

It turned out to be true, in one sense, that framing a shelter is essentially the process of sawing and nailing. It is also true, in another much more important sense, that framing is as creative as sculpting, needs the perspective of a muralist, and requires the owner-builder-construction crew to reach an agreement on just what should go where — before the process begins.

The principals at our place never arrived at such harmony; don't fret if you are also unable to think three-dimensionally and plan on just what sort of material and how much you will be needing for the next two weeks. These things work out; believe me, they do.

Once your foundation is committed, you have attained a certain error-limit. Your mistakes — and your successes — can not exceed the dimensional specifications of the posts, walls, slabs, or what-have-you that constitute the foundation.

We used those dimensions this way. First we put the great floor beams on the concrete foundation posts, ran the sills around the outer perimeter of the foundation, and then cut and nailed the floor joists across the beams. With that as a base, we made a subfloor of rough plywood sheets. The subfloor stayed until, on the very last day of interior construction, it was covered with the finish floor.

That subfloor was a matter of sawing and nailing.

It was also somewhat casual because we knew the finish floor had yet to come.

The outer walls and roof were something else again.

After we got the subfloor down, the place looked like a dance platform at a garden party; we danced around it in the evenings — it was so flat and so inviting, sitting there under the cool green trees.

That was one of our last joyous moments until the place was complete, some six months later.

"How high do you want the walls on the low side?" asked David Berry the next morning. David is my friend, still, but we had some tense moments during the construction. He managed the crew of inexperienced college-age and high school summer labor. We used sixty-three in all; none but six stayed for the entire project.

"I don't know," I answered. But the matter had to be resolved. The short-side post that went in on the northwest corner was a kind of keystone for the entire building. Every vertical dimension for each of the three sections of the building related directly to that post.

"Let's see," I mused, standing on the corner of the dance platform, my hand reaching over my head at an imaginary height. "I think it ought to be about here, about 8-1/2 feet."

"Okay," said Dave.

I left for the office for an hour or two of essential work. When I got back, the first frame (four posts and a beam) was up; the second was about to follow.

"It's too low," I said.

"I thought so," said Dave. The frame came down.

"That's too high," I said when I returned still later and looked at the second effort. "The living room front (high) side will be too high if we start so high back here."

The second effort came down. A day had gone by. The next morning a third version went up; it stayed; none of us was ever truly certain of the precise feet and inches. I have given "more or less" replies to dimensional questions ever since.

Most carpenters and builders will tell you that dimensions not only have to be precise, they cannot be set by the wave of a hand over the head. Surely the professionals have a point; but so do I — don't be dismayed if you can't project every last framing detail, like the height of the walls.

Once the post-and-beam framing went up, the roof (quite like the floor, in reverse) and the walls proceeded apace. I had made a cardboard scale model of the structure (something I advise for one and all) and had cut windows and doors where we wanted them (and, in most cases, needed them). From that, I had drawn a kind of floor plan. Looking at my renderings, the town's building inspector labeled me "the Grandma Moses of floor plan drawings." Using those sketches, plus the dimensions of the old doors we had rescued from a tottering, abandoned meetinghouse, plus the printed instructions on the doors and windows we had ordered from the lumberyard, the walls went up and the sheathing inside and outside went on over the proper amount of insulation. Almost all the doors and windows ended up in the places we had foreseen; there was some improvisation and adjustment along the way. Well, let's say there was a hell of a lot of improvisation, much of it damnably clever.

Somehow, whether or not you use a dodge like pretending houses are nothing more than boards and nails, what needs to get done gets done. You have to think about it, you have to look at it, you have to ask others to look at it; you have to read, you have to recall what you have seen in barns, which are the classic post-and-beam structures. You have to acknowledge that there will be mistakes and improvisations; but, at the same time, you have to proceed with the decisions that seem correct to you.

We designed and built our place around a couple of basic ideas. One was how the house would interrelate with the natural truths of the site; the other was how the people inside the place would interrelate with each other and the space available. In order to leave that decision as fluid as possible, we made no dimensional interior divisions as the floors, walls, and roof went on. The place was constructed like an airplane hangar.

After our framing was completed (and it was) we made the key decisions about the interior divisions.

How could we be so casual? How could we not have the kitchen in place, the boys' rooms outlined, the plumbing locations absolutely specified? I mean, other folks don't get the foundation down, the roof, the walls, and floor in place without knowing, or do they?

We could, and did; and, by our own measure, succeeded.

I believe that was possible because we built the place on some basic premises as well as some concrete posts. The north side of the house was built close (4 feet) to the woods because the woods would protect it from the cold, northerly winds of winter. The north side is also the low side of a basic, rectangular shed-roof structure because we wanted to minimize the area those winds could push at. Low and dark, that's the north side of the straight-line house that covered some 104 feet from west to east (leaving the long dimension facing north and south).

Low and dark. What do you think about when you hear those words? A kind of cave, perhaps? That's what came to our minds, and along that side (except for the main living room space, which straddles the entire north-south dimension) we put the bedrooms for five children. We also put a bathroom, a work-utility-storage room, entrance halls, and closets — each of which benefits from being located in the lower, darker space of the shelter.

On the south side, which is high, covered with windows, and open to the world through large, sliding glass doors, we put every major living and activity area — the places where the family, or parts of it, and guests gather to mingle, play games, read, socialize, converse. (Except for the so-called master bedroom — Jean and I both want to wake looking at the sun and the bay, and so we do.)

Because we knew why we wanted it to be the way it is, and were certain of our reasons, we also knew generally where the rooms would go. There was no need to sweat the specifics of their precise dimensions. Each bedroom on the north side got a small, ventilation window, installed in the general center of the room. The entire south side was covered with glass, so divisions and dimensions there did not matter. Ours is a window wall — a large one, which lets us live with the outdoors, indoors; and which admits the sun in the winter to heat us whenever it shines, the moon and the stars at night, and the breezes of summer whenever we need them.

When you consciously build in harmony with nature, you will find that your rooms place themselves, as do their designs. Because, for example, we believed everyone should share the sunlight, the brightness, the vitality of the trees and the view, we made certain every communal room opened to the south. That meant a minimum of internal walls. The kitchen, dining room, living room, and study are each part of, yet separate from, the other.

It is amazing how well it works, how the rooms function as secure spaces, with a kind of privacy and difference; and yet, at the same time, how they are an open part of the whole, allowing each person in each place to communicate with every other person in each of the other places.

I am continually astounded by how well it works. Others might be more astounded if they knew how unconsciously and ingenuously the effect was created. We never tried for it; we always tried for that natural harmony, a consciousness of the natural realities. That consciousness, and an uncompromising response to it, put our rooms in place, designed the kitchen (with some help from Jean), and positioned the doors and windows. A philosophy, if you will, created our internal space. None of it is wasted, none of it seems cramped.

It works, and the doors and windows work, because the house grew out of spiritual values rather than spatial science. We framed the place with only a casual concept of what was going to take place under the roof or within the walls, and we got away with it because it was the particulars of the dimensions that were casual, not the purpose of each interior space.

If you have some primary values in place and act on them, you too can be a Grandma Moses of floor plans and still get your shelter properly framed, with its doors and windows properly placed in relation to your interior divisions of space — as long as you don't forget just who it is who's living with you. There are nine of us altogether; and I must confess there were moments when I had provided for eight and a guest or two.

But that's because I can't add; not because we couldn't cope with the details of our own shelter.

(JNC)

17

Framing— Some Details

The frame is an integral part of each of the building surfaces. Being the part that gives most of the compressive and bending strength, it is analogous to the human skeleton. Many different framing systems have evolved, each satisfying some particular requirement of form, material, or function. Standard framing terminology is used in describing similar elements in all of these frames. Illustration 82 illustrates a large percentage of standard framing terminology.

Framing Systems

POST-AND-GIRT FRAMING

This method (see Illustration 83) was popular in this country until the early nineteenth century. In a time when trees were large and plentiful but power-sawed lumber was considered a luxury, the use of a minimum number of large hand-hewn beams resulted in the lowest building cost. Building well with these large beams required great skill. Every town had a framing man who would make mortise-and-tenon joints and supervise the raising. However, this type of decentralized operation was not compatible with mass production, so the building industry evolved the stick style of construction in which minimum-wage laborers put together houses in a way reminiscent of the Tinkertoy. Most people today assume that stick construction is cheaper than post and girt or post and beam. In reality, a properly engineered post-and-beam house of rough-sawed native wood costs about two-thirds as much in material as the standard stick house. Realization of this fact is giving the old post and girt a rebirth.

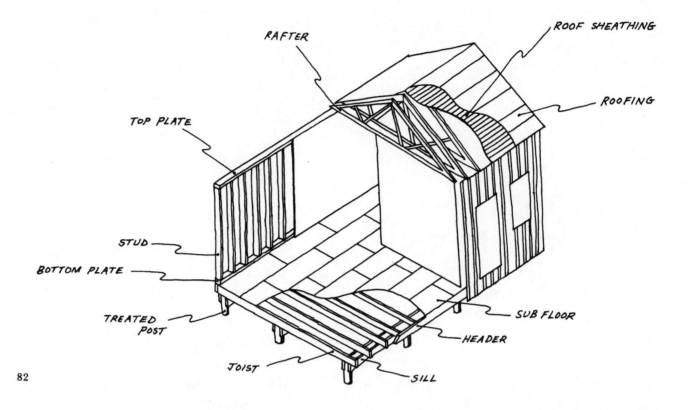

RAFTER
ROOF SHEATHING
ROOFING
TOP PLATE
STUD
BOTTOM PLATE
TREATED POST
JOIST
SILL
HEADER
SUB FLOOR

82

POST-AND-BEAM FRAMING

This is mainly distinguishable from post and girt in having no girts. The framing is still relatively heavy and of wide spacing. I usually make no distinction and call any framing system post and beam if the wall posts and roof beams or rafters are of 4-inch minimum dimension and the on-center-spacing is 4-foot minimum. There are several distinctly different approaches to post-and-beam construction. The most common is a double-pitched roof construction in which a large ridge beam carries one-half of the total roof load between posts in the end walls. A second dispenses with rafters entirely by the addition of more roof beams parallel to the ridge beam. A third is sometimes called *plank-and-beam* construction due to the use of 2-, 3-, or even 4-inch-thick planks to span considerable distances between rafters or beams. The fourth uses smaller beams (purlins) perpendicular to the rafters to limit the clear span of the 3/4-inch or 1-inch roof boards.

STICK-BUILT HOUSES

Around the turn of the century, the *balloon frame* (sometimes also called the eastern frame) was popular in multistory houses. The balloon frame is distinguished by having continuous studs from soleplate to top plate. Its popularity was due not to any cost saving but to the

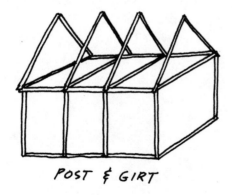

POST & GIRT

83

fact that the continuous stud resulted in the minimum frame shrinkage in seasoning. You'll recall from Chapter 11 that wood typically shrinks 0.1 percent in the direction of the grain but 5 percent across the grain in drying from the green state. In an era of plaster and stucco, this was very important.

STICK-BUILT (GORILLA CAGES)

84

However, long studs are now hard to come by and twice as expensive as 8-foot studs. And, instead of plaster, we now use more tolerant Sheetrock. A third fact, which had little influence on the industry but which we now recognize as important, is that the balloon frame as sometimes built was a natural tinderbox. The spaces between studs were potential chimneys through which fire could race between floors unless the firestopping had been installed with care.

The above reasons led to the now common *platform frame* (also called the western frame). A multistory platform-framed house is a series of single-story houses piled on top of each other. You can now see why "stud" has come to mean 2″ × 4″ × 8′. While this development led to the use of fewer types and dimensions of material, it also led to a subtle monotony in houses. Call today's houses by whatever term you wish (Cape Cod, split-level, garrison, ranch, etc.), they all have a built-in regularity that belies individual expression — the 8-foot stud.

THE ARCHES

Most roofs withstand vertical loading by bending strength. Masonry constructions, unless reinforced by steel, have virtually zero strength in tension and therefore zero strength in bending. The Roman arch overcame this limitation of stone by a geometric shape that caused each component to be purely in compression. The inherent stability of this design is proven by the existence of Roman aqueducts several thousand years old.

That remarkable substance, wood, however, has equal strengths in tension and compression. It is furthermore impossible to bend a piece with any depth into an arch without destroying the fiberbonds. Several framing methods have evolved in which wood has been used to approximate the arch to various degrees.

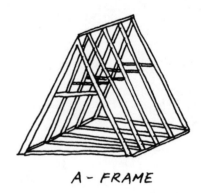

A - FRAME

85

The *A-frame* is the crudest approximation to an arch, being simply two straight pieces of wood joined at the top to function as a combined wall/roof. The single advantage of the A-frame is seen in considering it to be all roof. At the typical pitch of 24 (24-inch rise to 12-inch run), the snow load reduces to the minimum value of 20 psf. Therefore, the very long rafters can be quite small in section. The biggest disadvantage is seen in considering the A-frame as a wall. The inward sloping wall results in a large percentage loss of usable floor space. A further disadvantage is difficulty in heating. Hot air rises to the peak, making the peak too hot and the lowest level too cool. Adding a second story destroys the psychological effect of the cathedral ceiling.

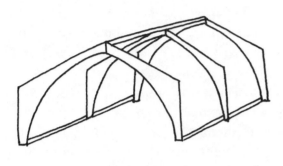

86   GOTHIC ARCH

The *Gothic arch* uses two glued-laminated arches that are structurally fastened at soleplate and peak. An extremely pretty arch, the Gothic is also very expensive, requiring fabrication under controlled factory conditions. While available for residential construction, it is used primarily in churches and public buildings where the great clear spans and height make it competitive with other systems.

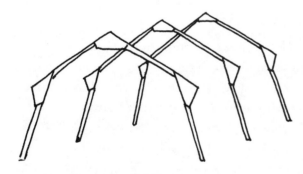

87   RIGID FRAME

The *rigid frame,* developed by the American Plywood Association, spans the greatest clear

volume using a four-straight-line approximation to an arch. The bending moments, caused by the straight-line deviations from a pure arch, are absorbed by plywood gussets that also conveniently connect the framing lumber. A variation on the optimal rigid frame allows the walls to be vertical at the expense of span. The rigid frame is second to no system in practically enclosing volume at minimum cost. An excellent rigid-frame handbook is available from the American Plywood Association.

Closing in the rigid frame requires minimum material and time. However, finishing the interior may prove extremely time-consuming if you don't like the appearance of the gussets. The frame is also not very flexible in placement of windows except in the end walls, which carry no load. The rigid frame as developed by the APA makes an excellent shop or barn. If its unique features appeal to you, you can also turn it into a remarkable shelter.

DOMES AND YURTS

A discussion of framing systems would not be complete without inclusion of the dome and the yurt. The *geodesic dome* is a fantastic engineering trick. Basically, it consists of a series of

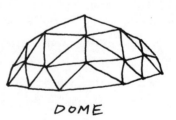

DOME   88

triangles whose sides (struts) are entirely in either tension or compression. Because it uses material in the absolute optimal terms — wood and metal are strongest in tension and compression; the only rigid geometric figure is the triangle — it is by definition the optimal use of framing materials. Being an approximation to the sphere, it also encloses the maximum volume using the minimum material. As in the rigid frame, the geodesic dome's interior space is entirely open. However, it is dangerous to focus on one spectacular aspect of a shelter and ignore the overall success. Walls and roofs have functions beyond optimal use of material. Problems with domes include:

(1) They are extremely difficult to seal against a driving rain.

(2) The sheathing is not square and so is materially wasteful and time-consuming.

(3) Like the A-frame, it is hard to heat without adding horizontal panels (floors), which then degrade the psychological effect of clear space.

(4) Acoustic privacy is impossible since the dome shape focuses sound.

(5) Doors and windows are difficult to incorporate.

(6) Insulating and interior sheathing is similar to exterior sheathing — wasteful and time-consuming.

YURT

89

The *yurt* is basically an assembly of identical wall panels and roof panels held together in compression against the floor and a roof compression ring and in tension against the hoop at juncture of wall and roof. It is neither remarkably inexpensive nor easy to build. Its validity lies in its original purpose — a nomadic equivalent to a mobile home. It is a pleasing and substantial wooden structure that, if planned

properly, can be assembled and disassembled effectively. I think, however, that the idea loses some of its appeal when considered as a permanent structure.

POLE HOUSES

Another example of barn turned inexpensive house is the "pole house." Deeply imbedded

POLE HOUSE

90

pressure-treated poles (telephone poles) do triple duty as foundation, posts, and bracing. Most areas of the country have individuals or companies equipped to auger the holes and set the poles. The average house would require about one day's time to set poles from start to finish. However, from that point on, the pole house becomes more difficult than the post and beam. Since the poles become part of the walls, placement of the poles is critical. It is very time-consuming to build a square house around round, tapered, and slightly crooked poles. If ordinary telephone poles end up inside the house, hot days will be aromatic of preservative for several years. If the poles remain outside the house you'll have, to say the least, a unique house — one that the bank mortgage board will have a hard time swallowing.

Don't confuse the pole house with the pole platform foundation, described in full under Foundations, in which the poles end at the sills. The conventional post-and-beam house built upon a pole foundation will actually take less time to build, and will have none of the above problems.

Functions of the Floor, Wall, and Roof

Most of the required functions of floors, walls, and roofs are identical. They all act primarily as membranes separating an inner controlled-environment space from the outside space. I repeat — before one chooses any aspect of a house he should:

(1) List all of the expected functions in order of priority.

(2) Use the list to evaluate the design options.

For example, my list of floor functions is:

to keep my feet dry;

to keep my feet warm;

to support a live load of 40 psf;

to exclude drafts;

to be level and smooth;

to be low in cost;

to be easy to build;

to wear well;

to look good;

to be easy to clean.

Wall functions include:

to hold up the roof;

to admit people;

to admit light;

to exclude winter cold and summer heat;

to admit summer breeze;

to resist wind pressure;

to prevent racking;

to be low in cost;

to be easy to build;

to look good inside and out;

to exclude bugs;

to provide a surface for storage and art.

Roof functions:

to keep out rain;

to support snow load;

to prevent winter heat loss and summer heat gain;

to resist wind uplift;

to be low in cost;

to be easy to build;

to shade south windows in summer;

to admit sun in winter;

to protect windows and doors from rain;

to be low in maintenance;

to look good inside.

There are dozens of different ways to build floors, walls, and roofs that satisfy different function lists. In the following illustrations, I will describe several options and their relative merits.

HOW TO BUILD A FLOOR

Illustration 91 shows the conventional floor and I don't like it because it requires a concrete foundation. I'll repeat what I said before about foundations: A crawl space house with insula-

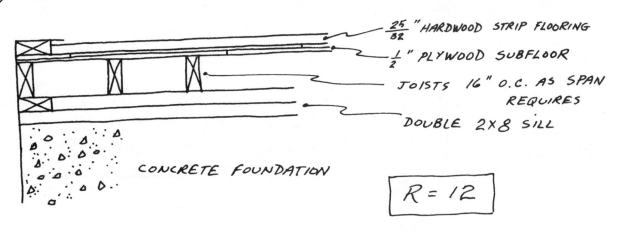

Ⓐ CONVENTIONAL

$\frac{25}{32}$" HARDWOOD STRIP FLOORING

$\frac{1}{2}$" PLYWOOD SUBFLOOR

JOISTS 16" O.C. AS SPAN REQUIRES

DOUBLE 2X8 SILL

CONCRETE FOUNDATION

R = 12

91

tion in the joist spaces has a warmer floor than the conventional full-foundation house without insulation. Also, I personally prefer a wide pine floor for beauty as well as cost.

Illustration 92 shows the "Rex Roberts floor." It is the ultimate in low cost but has a low R value (13), and the finish floor, serving also as the subfloor during the construction phase, suffers much abuse.

Illustration 93 shows the floor in my first house and still my favorite. It has a high R value (R = 26) and costs just a little more than B. A big advantage over B other than R value is accessibility. The joist space is available for wiring or plumbing by merely removing a single ledgerstrip. *Caution:* Do not substitute plastic film or builder's foil for the bottom panels unless clean holes are punched every 3 inches o.c. by a paper punch. Holes punched by a sharp instrument such as a knife or icepick do not provide adequate ventilation and will cause condensation in the insulation.

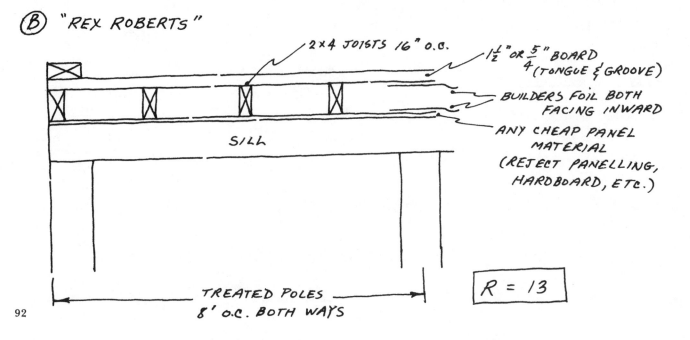

Ⓑ "REX ROBERTS"

2X4 JOISTS 16" O.C.

$1\frac{1}{2}$" OR $\frac{5}{4}$" BOARD (TONGUE & GROOVE)

BUILDERS FOIL BOTH FACING INWARD

ANY CHEAP PANEL MATERIAL (REJECT PANELLING, HARDBOARD, ETC.)

SILL

TREATED POLES 8' O.C. BOTH WAYS

R = 13

92

© MY FLOOR

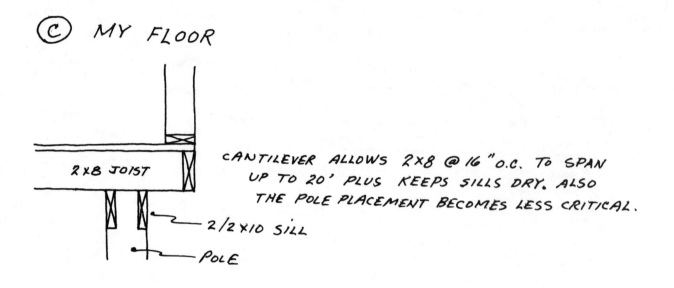

2×8 JOIST

CANTILEVER ALLOWS 2×8 @ 16" O.C. TO SPAN UP TO 20' PLUS KEEPS SILLS DRY. ALSO THE POLE PLACEMENT BECOMES LESS CRITICAL.

2/2×10 SILL

POLE

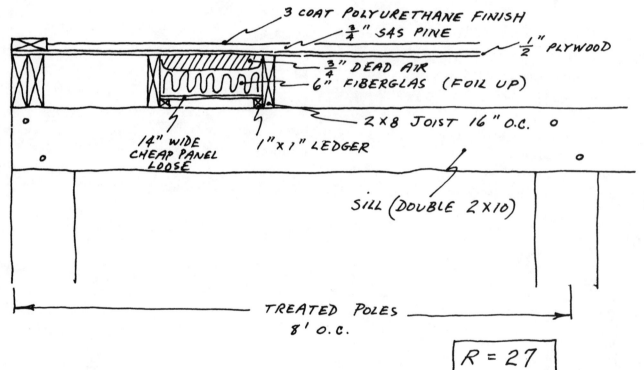

3 COAT POLYURETHANE FINISH

$\frac{3}{4}$" S4S PINE

$\frac{1}{2}$" PLYWOOD

$\frac{3}{4}$" DEAD AIR

6" FIBERGLAS (FOIL UP)

2×8 JOIST 16" O.C.

14" WIDE CHEAP PANEL LOOSE

1"×1" LEDGER

SILL (DOUBLE 2×10)

TREATED POLES 8' O.C.

R = 27

93

HOW TO BUILD A WALL

Few people realize that 1/2-inch CDX plywood can be as good a vapor barrier as foil-backed insulation as it is usually installed. If they do, they usually add a polyethylene vapor barrier over the foil, but then spoil it by cutting holes for electrical boxes. The use of 1-1/2″ × 3-1/2″ studs limits the achievable R value of this wall

pay is low R value (R = 9) and low racking resistance due to the absence of bracing.

The wall in Illustration 96 is essentially the same as Rex Roberts's, but horizontal stiffness is provided by shelving inside. Plywood is nailed to the shelving from the outside, using annular nails. This allows sheathing of 3/8-inch plywood, which prevents building racking, and 1-inch shelving spanning 48 inches. All dead

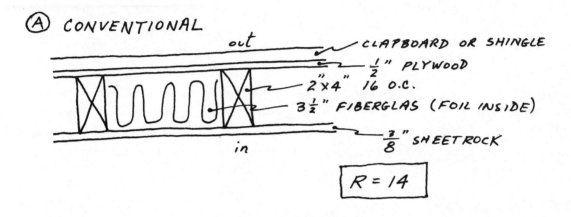

Ⓐ CONVENTIONAL

out — CLAPBOARD OR SHINGLE
½″ PLYWOOD
2″×4″ 16 O.C.
3½″ FIBERGLAS (FOIL INSIDE)
in — ⅜″ SHEETROCK

R = 14

94

to R = 14. The single advantage of the conventional wall is speed.

Once again, Rex Roberts has achieved the lowest-cost wall (Illustration 95). The interior wall is pleasing, the stud space is a natural for shelves, and the self-draining wall outside the sill line assures no sill rot. However, the price we

air/foil spaces are now self-draining. The R value is still low, but this makes an excellent shop or shed wall.

The wall in Illustration 97 costs no more than the conventional wall due to the use of rough framing and sheathing. However, both the thermal and vapor barrier performance are

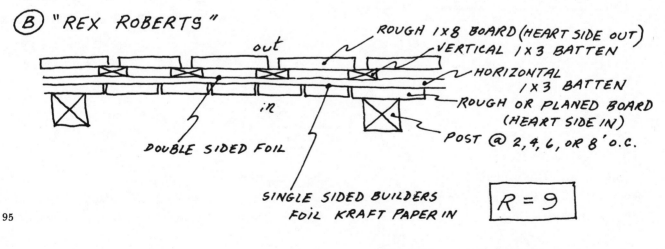

Ⓑ "REX ROBERTS"

out
ROUGH 1×8 BOARD (HEART SIDE OUT)
VERTICAL 1×3 BATTEN
HORIZONTAL 1×3 BATTEN
ROUGH OR PLANED BOARD (HEART SIDE IN)
POST @ 2, 4, 6, OR 8′ O.C.
in
DOUBLE SIDED FOIL
SINGLE SIDED BUILDERS FOIL KRAFT PAPER IN

R = 9

95

Ⓒ REX ROBERTS MODIFIED

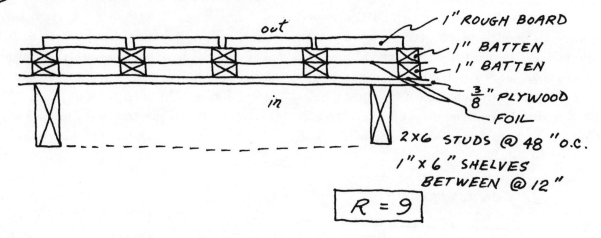

out

in

1" ROUGH BOARD
1" BATTEN
1" BATTEN
$\frac{3}{8}$" PLYWOOD
FOIL
2×6 STUDS @ 48" O.C.
1"×6" SHELVES BETWEEN @ 12"

R = 9

96

superior, not to mention the beauty and maintenance factor. The R value is extremely high (R = 26). The 1" × 3" horizontal strapping provides a free air space within which electric

sheathing cutouts without penetrating the vapor barrier. The vapor barrier is installed as a single sheet of black polyethylene strapped around its perimeter like a picture frame, there-

Ⓓ MY FAVORITE HOUSE WALL

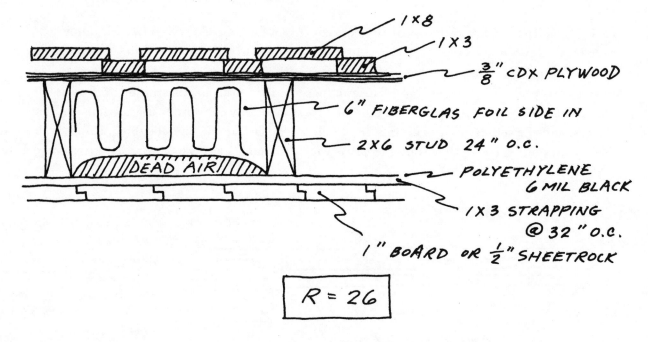

1×8
1×3
$\frac{3}{8}$" CDX PLYWOOD
6" FIBERGLAS FOIL SIDE IN
2×6 STUD 24" O.C.
POLYETHYLENE 6 MIL BLACK
1×3 STRAPPING @ 32" O.C.
1" BOARD OR $\frac{1}{2}$" SHEETROCK

DEAD AIR

R = 26

97

wiring is run. The cable (NM, UF, or conduit, according to local code) is run along the strapping, and outlet boxes are mounted in wall

by completely eliminating water vapor condensation within the insulated wall space.

The choice of exterior wall surface is yours.

The old-timers often used cedar shingles on the north wall, due to their resistance to driving rain and any form of rot, but used vertical pine boards on the sunlit sides for long life and low cost. My personal choice is reverse board and batten consisting of 1″ × 8″ rough-sawed pine boards and 3/4″ × 2-1/2″ S2S ("strapping") battens. No stain or preservative is required ordinarily. Don't forget to season and nail the boards *heart side out*. Reject boards with excessive twist due to cross grain. Reverse board and batten is unusual and handsome and the easiest siding to install. Most important, however, is its weatherability. If the boards ever get wet in a driving rain they dry rapidly, allowing little chance for a fungus to attack, due to the air space between the plywood sheathing and the

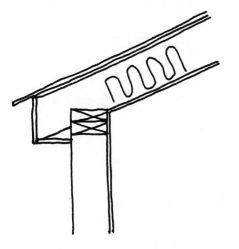

boards. In areas of high vapor pressure (bathroom, kitchen, and laundry), ventilation holes can be drilled in the plywood at the top and the bottom of each stud space and never be seen under the boards.

HOW TO BUILD A ROOF

The conventional roof (Illustration 98), like most conventional surfaces, suffers from a singular lack of character. Its antiseptic white, flat surface assures us of nothing when the winds blow and the rains drive. The first indication we get of trouble is a yellow stain on the white-white ceiling. We usually cover it with asphalt shingles, never thinking that we are paying extra for that shinglelike tab and getting less than one-half the protection of double coverage roofing.

Rex Roberts has preserved the integrity and beauty of the roof by allowing the rafters and purlins to show, just as in an old-time barn. This roof (Illustration 99) costs the same per square

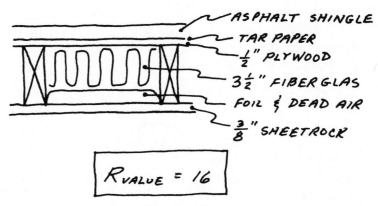

98

foot as a conventional roof. However, the insulation value of dead air/foil space depends upon the orientation of the space and the direction of the heat flow (see Chapter 19). Rex Roberts and I part company around January in Maine!

The roof in Illustration 100 (page 151) is my favorite for the following reasons:

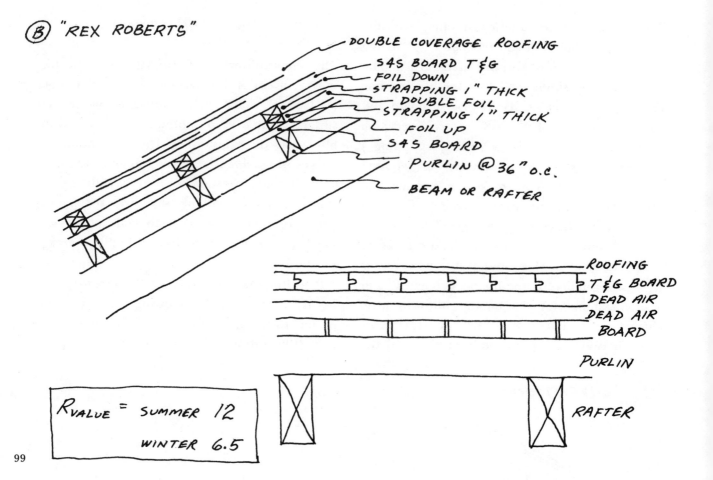

(B) "REX ROBERTS"

DOUBLE COVERAGE ROOFING
S4S BOARD T&G
FOIL DOWN
STRAPPING 1" THICK
DOUBLE FOIL
STRAPPING 1" THICK
FOIL UP
S4S BOARD
PURLIN @ 36" O.C.
BEAM OR RAFTER

ROOFING
T&G BOARD
DEAD AIR
DEAD AIR
BOARD

PURLIN

RAFTER

R_{VALUE} = SUMMER 12
WINTER 6.5

99

(1) it has an exposed ceiling of heavy beams and pine roofers (purlins can be used if you wish);

(2) it has a large R value (R = 30);

(3) the living space is easily sealed by the plywood sheathing since the rafters end at the walls;

(4) the rafter is cheaper because it is shorter;

(5) the double overhang protects both walls and windows and doors from rain;

(6) the overhang balances snow loads so that a smaller rafter section is permitted;

(7) rigidity is transferred from rear and side walls to the south wall by the plywood roof diaphragm; and

(8) all work is performed from above and so it is even faster to build than a conventional roof.

The bottom rafter section modulus is calculated from Case 5, Table 3. If the front and rear overhangs are not equal, use the average length for a in the formula. The cantilevered snow load relieves the maximum bending moment at the bottom rafter midpoint, often enough to reduce the required size of timber significantly. If the cantilever exceeds 4 feet, the bending moment in the top rafter at the outside wall line should be calculated using Case 3, Table 3.

The upward wind pressure against the overhang of 30 psf requires special care in fastening the top rafters to the wall. Plywood sheathing acts as a gusset in fastening the bottom rafters, top plate, and wall posts together. Therefore, fastening the top rafters to any one of these is equivalent. The simplest method I have seen consists of nailing 1" × 12" × 16-gauge predrilled galvanized straps, one on each side, to the top rafters and plywood under the exterior finish material. Commercial versions of the same device may be available with the trade names Du-Al-Clips, Ty-Down-Anchors, or Trip-L-Grips. Final fastening must be delayed until the rafters have completely dried.

As an option, purlins can be run over and at right angles to the bottom rafters, allowing the tongue-and-groove roofers to run parallel to the slope. The top rafters are then toenailed through the roofers into the purlins, a more secure arrangement. I prefer 3" × 4" or 4" × 5" purlins.

The vapor barrier should ideally be a single sheet of black polyethylene (manufactured in 100-foot rolls up to 40 feet wide), installed immediately above the tongue-and-groove roofers and below the top rafters. If the vapor barrier is damaged during construction of the roof, seal the damaged area using silicone building sealant. Don't worry about the top rafter toenails penetrating the vapor barrier. They completely fill their holes.

The most convenient and cheapest insulation to use for this roof is 4" × 23" × 32' Reinsul, a roll fiberglass without backing. Two layers are used to achieve an 8-inch thickness between 2" × 10" rafters. When closing the space between top rafters, leave a 2-inch gap at the top, at both the front and rear of the roof, as a ventilator. Unless you're into housing red and flying squirrels, screen the opening with 1/4-inch galvanized mesh.

The front and rear overhangs protect the wall sheathing, windows, and doors from rain, thus prolonging their lives. For the same reason side wall overhangs should be provided. A 12-inch side overhang is easily provided by extending either the roofers and/or the plywood an extra foot beyond the wall line and finishing off with a fascia board. Sometimes the overhang will look better with nonfunctional buttresses from wall to fascia at 8-foot intervals.

The edge of the plywood is exposed to extreme weathering conditions. Either Pentadip this edge every few years or cover it with a drip edge.

My favorite roofing material is white double-coverage roll roofing (selvage) with a 19-inch overlap. Don't confuse this material with 90-

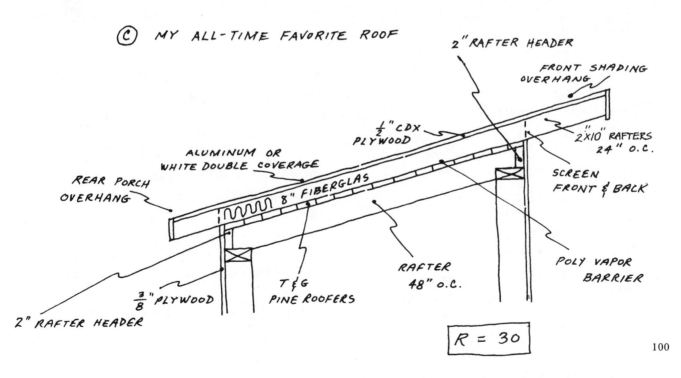

© MY ALL-TIME FAVORITE ROOF

FRAMING — SOME DETAILS / 151

pound mineral-surfaced roll roofing with a 3-inch overlap. The former is a good twenty-year roof while the latter is intended for temporary service only. Selvage comes in 36-foot-long rolls and, provided it is installed in warm weather, it can cover a 36-foot-wide roof in one continuous strip, manufacturer's instructions notwithstanding. Selvage can be installed on a roof with slope as low as a 2-inch rise in a 12-inch run. Asphalt shingles should not be installed on a slope lower than 5 inches in 12 inches. My favorite roof slope is either 3 inches or 4 inches in 12 inches.

Doors and Windows

HOW TO THINK ABOUT DOORS

As with other parts of the house, the way to begin thinking about doors is to compile a list of their functions. We will find several necessary functions and a number of secondary functions. Generally speaking, the more secondary functions, the worse the primary performance. I expect a howl of protest, but, in my experience, the sliding glass door provides a good example of poor primary performance caused by a secondary function. My door function list is:

(1) to let people in and out easily;

(2) to let large objects in and out;

(3) to keep winter heat in;

(4) to keep winter wind out;

(5) to keep water out;

(6) to let summer breeze in;

(7) to keep summer bugs out.

Functions 1 and 2 are primary. Numbers 3, 4, and 5 are also primary functions of a wall. Numbers 6 and 7 would be nice but could be provided by other means. In fact, a hinged section of wall would perform functions 1 through 5 very well and the addition of a screen door or screened entranceway will take care of 6 and 7.

Before we design our door let's list problems we have all had with store-bought doors and specifications for our door.

Problems with Conventional Doors:

Swelling and Sticking. The conventional exterior door is made up of lumber with grain running in both directions. If it gets wet it swells 1/4 inch or more in both directions. But we have to live with the sticking door because we know any material we remove will become daylight through which the winds will blow next winter.

Infiltration. The average house probably loses a third of its heat through infiltration or physical replacement of warm air by cold outside air through cracks around doors and windows. Effective weather-stripping of conventional doors is virtually impossible because of the way they close and their dimensional instability.

Hanging. Hanging a conventional door is not a job for the amateur. The opening tolerances that must be held for any degree of success are more suited to the cabinetmaker than the first-time home builder.

Weathering. The better-looking doors are made by gluing thin panels into a thicker framework like a piece of furniture. How long do you think a piece of furniture would maintain its integrity if a child slammed it against the wall every day? The panels soon loosen in the frame and then even good paint can't keep water out of the joints.

Specifications for Our Door:

Size. At least one outside door 3'0" × 6'8".

Strength. Strong enough to withstand children hanging from it, head-on collisions with Tonka trucks, armloads of firewood, and larcenous intruders.

Thermal. Insulated and sealed like a refrigerator door.

Convenience. No swelling or sticking (no planing ever), no painting, no replacement of hinges.

Cost. Cheap.

HOW TO BUILD THE DOOR

First cut a piece of 3/8-inch or 1/2-inch *exterior* plywood with one smooth side. If you use

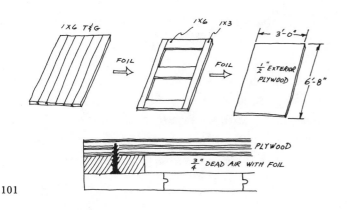

101

1/2-inch CDX for sheathing your roof, you'll probably find several pieces that could be used to save money. Next glue and screw a frame of S4S boards around the perimeter (1" × 3" verticals and 1" × 6" horizontals). The center horizontal is installed at latch level and provides a solid core for installation of the latch and middle hinge. Apply a layer of builder's foil to the inside of the plywood using contact cement or lots of staples. Cover the framework with another layer of foil. Finally, screw or

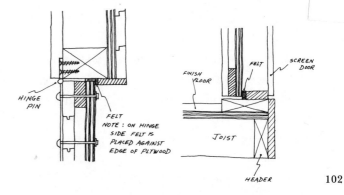

102

toenail matched tongue-and-groove roofers or shiplapped rough pine boards vertically, horizontally, or diagonally in any pattern that pleases you. Options include using S4S 2" × 4" verticals and 2" × 6" horizontals and filling the space with foam insulation.

HOW TO HANG THE DOOR

When you built your wall you provided two heavy posts (say 4" × 6") 36-1/4 inches apart to serve as the door frame. The bottom of the frame is the wall soleplate and the top of the frame is a 2" × 6" nailed horizontally 6'8-1/4" above the soleplate. The plywood wall sheathing is nailed over the frame flush to the door opening. Infiltration around the frame is eliminated and the door frame is extremely rigid and strong.

Next install three large T hinges with their tongues over the top, middle, and bottom 1" × 6"s, using carriage or machine bolts of sufficient length to pass clear through the door. Place the door in its opening on a 1/4-inch-thick shim. There should be about 1/8-inch clearance on both sides and none at the top. Trim the door if necessary. Fasten the hinges to one of the house frame posts using 2-inch wood screws. When the shim is removed, the door will swing freely with 1/8-inch clearance all around. Next, install the thumb latch or other door-closing hardware.

We have saved the best for last. Here is what will make your door as tight as a refrigerator door. With the door latched, nail wood doorstops all around — top, sides, and bottom — right up against the door. Then, using contact cement and staples, apply old-fashioned felt weather-stripping to the doorstop so that the door face closes against the felt face. You'll have to lean into the door to latch it, but in so doing you will be compressing the felt against the flat surface of the plywood and never a breath of winter air will infiltrate that door (see Illustration 102)!

A few fine points: You have a choice of whether the door opens inward or outward. If outward, a screen door can be installed on the inside. If inward, you should protect the entrance against the rain (you should anyway). Both of these functions are satisfied by an entranceway, which I favor and will describe next.

Compare this homemade door to the commercial variety. It costs less than a third as much; is insulated at least three times better; has a positive refrigerator-type seal; will never sag or fall apart; can't swell and stick; is easy to hang; is stronger; looks good to you.

SALVAGED DOORS

If you start looking a year ahead of time, you can find doors salvaged from mansions, banks, post offices, etc., at about 10 percent of their replacement cost. These rather funky objects can add something to your house impossible to duplicate with new material. However, the place for such doors is generally between interior spaces not facing the weather. Bank doors are really different only in being larger, heavier, and of more exotic wood. Therefore, the homemade door I just described will outperform them in

every way as an exterior door. When it comes to interior doors, which are expected mainly to isolate interior spaces acoustically, even a stained-glass window on hinges is workable.

An Entranceway

For a number of reasons I like a house built as a group of modules. One of the benefits of such an arrangement is the unique entrance it allows. Illustration 103 shows the entrance to my house. A shed-shaped passageway is suspended between the two houses. The back and roof are insulated with fiberglass; the bottom consists of spaced 2″ × 6″s on face with plastic foam insulation beneath. The front wall is covered with Plexiglas panels in the winter and screen in the summer. This entrance functions as an air, water, and mud lock. Shoes, coats, mud, and water are left behind in this space, keeping the rest of the house clean and uncluttered. People coming to the door may be

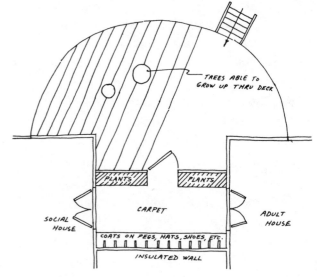

103

invited no farther than this space without being overly offended. When it's raining or cold outside I can say, "Come in," and still retain the option of not inviting them any farther. The first door is closed before the inner doors are opened just as in a submarine air lock. By this means, air infiltration is greatly decreased.

If one didn't choose to build in modules, the same effect could be achieved by entering through a sun-heated greenhouse. My entrance-way is 5 feet wide. The next time, I'll make it 8 feet to allow more room for plants.

Windows

HOW TO THINK ABOUT WINDOWS

The first windows were used to admit light to dark, gloomy places. The development of sheet-glass drawing allowed one to "see" an event on the other side. Techniques of sheet-glass production have improved to the point where today the material cost of a window wall or wall consisting almost entirely of glass is only about twice that of a conventional wall. The price of ordinary glass has not increased significantly in twenty years and is now a building bargain. Whereas a few decades ago the window-wall house was a luxury reserved for the rich, today, heating costs included, the solar-tempered house is cheaper.

My function list for windows is:

	(1) to admit light;
most impor-tant	(2) to bring the outside in (more than merely seeing out);
	(3) to keep winter wind out;

| impor-tant | (4) to keep rain out; |
| | (5) to keep winter heat in; |

| least impor-tant | (6) to let summer breezes in; |
| | (7) to keep bugs out. |

I have never personally owned a window (and I've bought all of the famous brands) that performed 6 and 7 without to some degree infringing upon 3, 4, and 5. On a per-square-foot basis the ability to open a window just about doubles its cost. The area of opening required for ventilation is about 10 percent of the area desired for transmission of light.

The case for fixed, nonopening windows is so strong I feel no need to pursue it further. Fixed windows can also be purchased from the local lumberyard. Should we buy or build? Consider the window frame and its function. Is not its function to facilitate the fastening of the glass to the wall? When we buy a fixed window we are, in one way of thinking, buying a piece of glass in a fancy shipping crate, which we then fasten to the wall frame. That is, we fasten a frame to a frame. There is never any infiltration between the glass and frame but only between frame and wall! Therefore, let's eliminate the middleman's expensive frame and seal glass directly to the wall for a totally infiltration-free window!

Window Geometry: There should be purpose as well as beauty in our window-wall geometry. The geometric considerations are shown in Illustration 104. The window should extend to a height such that it is fully illuminated at noon on the date of the lowest sun angle, December 22. The combination of window bottom height and overhang should result in zero direct sun on the date of the highest sun, June 22. Most people prefer a low windowsill so that they can see out

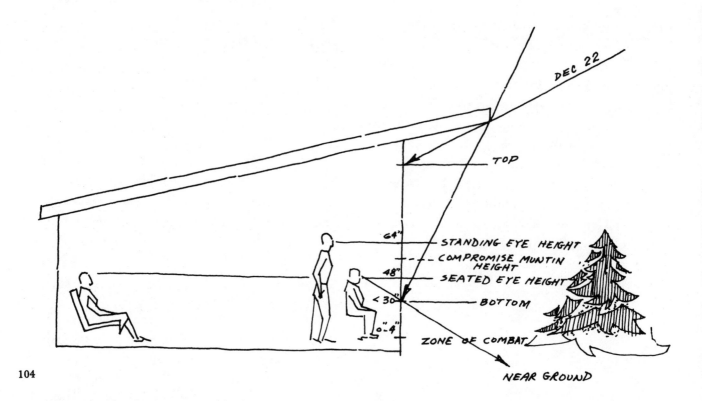

without craning the neck while seated. In case you're tempted to extend the glass all the way to the floor, however, remember that the baseboard is a regular combat zone, having to repulse balls, Tonka trucks, and vacuum cleaners on casters. At the very least give the window 4 inches of protection.

While the eye is capable of ignoring a muntin at a distance, any obstruction in the field of view at close range is distracting. Therefore, avoid the height of the average seated (48 inches) and standing (64 inches) eye for your horizontal dividers.

Double vs. Single Glazing: You'll find in Chapter 19 that double glazing is justified in terms of decreased fuel costs for all but the southernmost states. Do I disagree with Rex Roberts? I don't think so. Roberts wrote a decade ago. Heating costs have tripled in that period while glass prices have remained fixed. I'm positive that he would now agree with my

modification of his insulation standards. What about triple glazing? Yes, but the more glass we add, the more we confuse the condensation issue. I draw the line at two layers of glass but recommend a metalized mylar (available commercially as Astrolon or "space blanket") reflective night shade, equivalent in insulation value to three additional layers of glass.

We still have to make the choice between factory-sealed insulating glass and installing two separate sheets. Your decision will depend upon the dollar value placed upon absolute freedom from condensation. Commercial insulating glass costs $3 to $4 per square foot and will fog on the living side surface only on the very coldest days. The owner-built double glazing will cost $1 to $1.50 per square foot and will fog lightly over a quarter to a half of its between-pane surfaces on winter days with the wind out of the north. It will also require cleaning between panes every few years.

HOW TO INSTALL COMMERCIAL INSULATING GLASS

Illustration 105 shows the installation of a factory-sealed unit in a homemade frame. The top and bottom frame members are made from 2-inch pine stock on a bench saw. Notice that the window backstops are integral, preventing leaking. The side stops are strips of wood nailed to the wall posts or side frame members with a nonhardening building sealant between to prevent leaks. Silicone building sealant is applied generously all around the stops. Neoprene setting blocks are snapped onto the window units a quarter of the way in from all corners of the

window unit. The unit is then pressed into place, bottom first, in the compound. Last, a face stop is nailed all around. All of the supplies except the wood can be obtained through your glass dealer. Before installation, soak all pine frame pieces overnight in a clear preservative. A repeat application about every few years will keep the frame eternally new.

Table 9 lists the standard sizes of commercial insulating glass. Don't forget to make the window space 1/2 inch wider and higher than the glass.

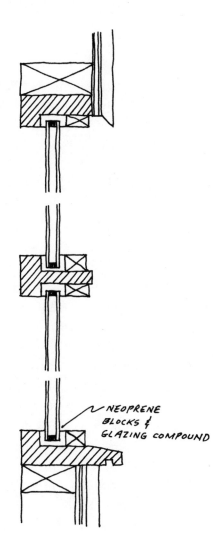

NEOPRENE BLOCKS & GLAZING COMPOUND

105

Table 9. *Standard Sizes of Insulating Glass (LOF)*

*2 panes of 1/4" polished Parallel-O-Plate
1/2" air space. 6.5 lbs. per sq. ft.*

Width		Height	Width		Height
33"	X	76-3/4"	64-1/2"	X	50"
35-1/2"	X	36"	64-1/2"	X	58"
35-1/2"	X	48-1/8"	64-1/2"	X	66"
35-1/2"	X	60-3/8"	66-5/8"	X	47-3/4"
42"	X	66"	66-5/8"	X	60-1/8"
42"	X	72"	68-3/4"	X	36"
44-1/2"	X	36"	68-3/4"	X	48-1/8"
44-1/2"	X	46"	68-3/4"	X	60-3/8"
44-1/2"	X	48-1/8"	68-3/4"	X	72-3/4"
44-1/2"	X	60-3/8"	70-1/8"	X	52-1/2"
44-1/2"	X	72-3/4"	70-1/8"	X	56-1/2"
45"	X	76-3/4"	72"	X	48"
45-3/8"	X	52"	72"	X	60"
46-1/8"	X	52-1/2"	72-1/2"	X	46"
47-3/4"	X	50-3/8"	72-1/2"	X	50"
48"	X	48"	72-1/2"	X	58"
48"	X	60"	72-1/2"	X	66"
48-1/2"	X	42"	75"	X	36"
48-1/2"	X	46"	75"	X	48-1/8"
48-1/2"	X	50"	75"	X	60-3/8"
48-1/2"	X	58"	80-1/2"	X	50"
50-3/8"	X	47-3/4"	80-1/2"	X	58"
50-3/8"	X	60-1/8"	84"	X	66"
55-1/4"	X	36"	84"	X	72"
55-1/4"	X	48-1/8"	93"	X	36"
55-1/4"	X	60-3/8"	93"	X	48-1/8"
56-1/2"	X	42"	93"	X	60-3/8"
56-1/2"	X	46-1/8"	93"	X	72-3/4"
56-1/2"	X	50"	96"	X	66"
56-1/2"	X	58-1/8"	96"	X	72"
56-1/2"	X	66"	96-1/2"	X	50"
57"	X	76-3/4"	96-1/2"	X	58"
58-1/8"	X	52-1/2"	116-1/2"	X	58"
64-1/2"	X	46"			

Unfortunately, none of the standard sizes of insulating glass are an ideal width for 4-foot-on-center post-and-beam construction either with rough 4" × 6" or S4S 4" × 6" posts. We can either have the posts sawed specially, use 2" × 6" or 6" × 6" posts, or order the glass specially.

HOW TO BUILD YOUR OWN INSULATING WINDOWS

Illustration 106 shows the detail of a homemade double-glazed window. The window framing required is not much more difficult than for the commercial units. These windows are remarkably free of condensation.

Air has the ability to hold a certain percentage of water vapor. When the relative humidity of air reaches 100 percent, the air can hold no more and the water vapor begins to come out of the air as condensation. The warmer the air, the more water vapor it can hold. If cold winter air moves from outside into the space between the windows it can only warm up and increase its ability to hold water vapor. On the other hand, air from inside the house will fall in temperature and ability to hold water vapor as it moves outward. If the relative humidity of the inside air and the temperature drop are sufficient, water will condense on the inside of the outside pane. *Therefore, if we could guarantee that the air between panes originated from the outside we would never have condensation!* Can we guarantee this? Nearly. We install the inside glass with a heavy application of silicone building sealant between stop and glass and between stop and frame. The outside stop, on the other hand, is installed very loosely with a felt weatherstrip between wood and glass. The weatherstrip 1) allows ventilation of the air space from outside, 2) filters out dust that

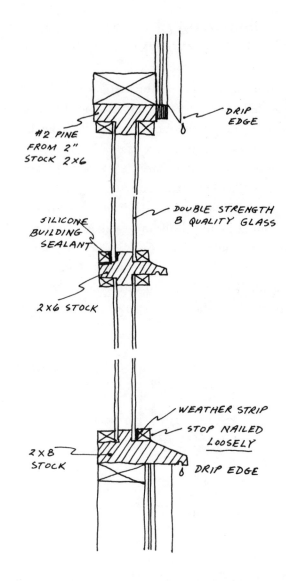

DRIP EDGE

#2 PINE FROM 2" STOCK 2X6

DOUBLE STRENGTH B QUALITY GLASS

SILICONE BUILDING SEALANT

2X6 STOCK

WEATHER STRIP

STOP NAILED LOOSELY

2 X 8 STOCK

DRIP EDGE

106

otherwise might require window washing every year, and 3) cushions the glass to prevent the glass rattling against the frame in a wind. The only time these windows will fog is when the wind blows from the opposite side of the building (a north wind for south-facing windows). A low-pressure suction is produced on the downwind side, pulling air from the inside.

While it is impossible to guarantee freedom from condensation, these windows are better insulators than commercial units. The full 1-inch air space more than compensates thermally for the slight flushing action. I have four small commercial Thermopane awning windows, used for ventilation, which consistently frost on the inside before any of my homemade windows.

Warning: In the rush of your enthusiasm, do not shortchange yourself by neglecting five very important points: (1) Use *pine, cedar, cypress,* or *redwood* for the window framing. Other species will not last as long. (2) Slope the sill as shown to dispose of standing water. (3) Treat sills and muntins with a clear wood preservative, such as Pentadip or Cuprinol. (4) Install outside glass first, and allow a few months of heated winter occupancy before installing inner panes. Otherwise the shrinkage of the wall framing will stress the silicone seal. In addition, the water from the drying frame will be deposited between panes as visible condensation. (5) Install outside wood stop with screws or small nails with heads projecting to facilitate removal of outside pane for occasional washing.

BUYING GLASS

Standard sizes of common window glass include every combination of even inches between 6″ × 8″ and 60″ × 70″. You should frame the window 1/2 inch wider than the glass and 1/4 inch higher. A perfect combination would be using S4S or S2S 4″ × 6″ (3-1/2″ × 5-1/2″ or 3-1/2″ × 6″) 4-foot-on-center posts and 44-inch-wide glass. The only trouble-free way to frame a window wall is flat on the floor deck using identical 44-1/2-inch-long muntins (horizontal frame members) to space the posts. Any post not perfectly straight will be forced straight by the muntins acting against adjacent posts. After the wall is raised and braced, you can feel confident that the glass will fit on the appointed day.

Glass by the case is discounted 50 percent. The price of glass is nearly constant up to a size of 18″ × 18″.

Trapezoidal windows can be installed at virtually no increase in cost but with a big boost in class! Glass suppliers will often cut glass at no charge if you order enough. One cut of a standard piece of glass results in two trapezoids, either identical or of different size, depending upon the location of the cut.

One final word of caution. Don't order glass ahead of time. It can be obtained in any quantity within a week at the most. I've found it impossible to store glass safely at a building site. I installed more than a hundred pieces of glass in my house — some as large as 44″ × 72″ — and broke only two in installation. But storage losses ran to over two dozen.

(CW)

18

Heating

My first law of heating: The more precision and craftsmanship you have used to construct your floors, walls, roof, doors, and windows, the easier (and less costly) it will be to heat your home.

My second law of heating: Whatever ingenuity you can use to cut your dependence on external energy sources (the heating oil truck, the electric company, or the bottled gas delivery system), the more you will glow with pride; and thus, the warmer you will feel.

I have never lived in a warm climate. Except for a month or so in Florida (courtesy of the U.S. Army Air Force), my fifty-three winters have all been cold ones.

Those that were spent in dormitories, steam-heated apartments, etc., have been well to excessively heated.

Those that have been spent in at least ten different wooden frame houses have been drafty and expensive. The problem of heating the average rural home once the temperature drops below freezing and the wind climbs to above 25 mph is a problem very few persons have solved unless: 1) they are willing to spend inordinate amounts of money for heating fuel; or 2) they begin to exercise their cerebrum and think the problem through.

I did some of that work before we built our post-industrial house and I've done some since. Here, in outline form, are some of the points I've covered:

• Tightness. There is no describing the feeling of warmth that a tight house can give. I have only been in one. My mother used to feel cold when the temperature dropped below 72° F. She could feel a draft that would not flicker a candle. She announced these propensities to the contractor who built a small home for her in Maine in the early sixties. She announced them each day.

Like so many persons who knew my mother, the contractor must have been in awe of her intensity. At any rate, that $35,000 home was the tightest I have ever experienced. If a cigar smoker could have produced a smoke ring on

Friday, left the living room slowly, and eased the door shut until Monday, my mother would have seen the ring in the same place when she reopened the door after her weekend holiday, which had been spoiled by a hurricane. And she would have been cross, because she could never stand cigar smoke.

They couldn't build the older homes as tightly; they didn't have the insulating materials, the fastening techniques, the double-pane windows. Using relatively unskilled or merely enthusiastic help to build your new shelter, you will probably not be able to build as tightly either. It takes an awesome amount of time and precise work, both of which, I discovered, are difficult to come by when you are in a hurry to complete your new home — especially if it's the first you've built.

As a result, you'll probably do what we've been doing each winter since we moved in. We've been caulking, covering, weatherstripping, putting up molding over seams, banking the house exterior with straw, and generally locating and blocking every possible spot where outside air can sneak in, or vice versa.

That's quite a job when the wind blows from each quarter at something more than 60 mph during some part of the winter. Which brings up another point:

• Thinking heat. The problem with most folks is their acceptance of drafty places, high fuel prices, and the general discontent that both can sponsor. If more of you fought against discomfort and for independence and economy, you would find ways to keep warmer for less money.

Take the winter wind, for example. Whenever it blows across the land north of the Mason-Dixon line, it robs homes of heat. Why should not the opposite be true? With a windmill, designed to feed the electrical energy it can generate directly into household heating units,

the harder the wind blows, the warmer the house will get.

Such wind dynamos are abuilding, and I plan to get one. But, by applying the same principles that led me to the windmill idea, I have already made the house warmer and costs lower.

The most obvious and basic step is the installation of large, double-pane windows along the entire south side of our home. Fortunately (and this process begins before the site is selected), that south side faces Middle Bay. The view is splendid; no one can peer in the huge windows; when there is no snow on the ground, the bay acts as a reflector of solar energy; when there is snow, the sun's work is multiplied.

With some 2,800 square feet of living space in the house, we save about 40 percent of the fuel oil compared to what we burned previously in our smaller, more conventional, restored Maine farmhouse, built in the 1870s. The irony, of course, is that no sooner did the sun help us begin saving than the price of fuel oil went up more than 40 percent.

That further prodded me into lowering the thermostats and keeping two wood fireplaces going most of the time. The woodburners are ceramic firepots, to be exact, and they function quite well, although not as efficiently as cast iron wood-burning stoves. (We are pondering the addition of one of those. Wood comes easily to us. We are surrounded by woods on the land side of the property, and the bay brings us driftwood year-round.)

Still, at 1975 prices, we have to spend about $600 a year for fuel oil used in the hot-air furnace that's under the floor. When the windmill arrives, that $600 will be cut by more than half. With further tightening, that $300 could be cut by $100. With the installation of one more wood stove (and the persistence to use it regularly), the final $200 could be cut to $100 — which is more like it. Then, if we wanted to get reflecting/insulating curtains to pull across the windows at night, or on sunless days, we could do even better.

By lowering our "comfort zones" just a bit and adjusting to them, the eight of us that live in the house could be content, well warmed, and quite comfortably and spaciously housed for about ten bucks a person per winter. (The wind dynamo, by the way, is being fitted with a device that allows it to pump surplus electricity back into the power lines. That turns the meter backward, and, if the wind blows, the power company could end up owing you money. If that dream could be realized, the heat of my enthusiasm would cut the oil bill even further.)

There is much to be gained by thinking heat. The same technology that produced the insulation materials that make heating efficiency so possible has also spawned the mass-production, mass-consumption, mass-waste systems that are depleting the nonrenewable heating resources that have been taken for granted for more than a century. It makes eminent sense to use the new insulators to conserve what's left of coal, oil, and gas. It also makes sense to ponder ways to cut the consumption of such nonrenewables.

They are, after all, the last apples in the barrel; and the apple sellers who control the markets are going to charge more and more as the bottom of the barrel begins to show. It is certain that as each year goes by you will be paying more for any nonrenewable heat source, whether it be electric heat generated at a nuclear plant or fuel oil carried from Arabia, natural gas from Texas, or coal from West Virginia. As the price of one goes up, so will the price of the others, as will the profits of the sellers of the last apples.

The solution is a change of your energy diet. Switch from nonrenewables to renewables, and use the best of what technology hath wrought to help you. Put in as much insulation as you can afford, or as much as will be efficient. Build into the sun, away from the winter winds. Don't count on gadgets more complicated than windows on the south, windmills that produce heat directly (without expensive storage systems), and wood stoves that burn efficiently, especially if you live near a source of wood supply.

Add tightness and conservation to these, and you'll have a place even my mother could love. You'll also be denying the energy exploiters a source of windfall profits, and that should give you pride.

And, finally, you'll be running your household heating system the way all such systems are going to have to be run a generation or so in the future: with renewable energy resources locally produced.

You should feel warmer already.

(JNC)

19

Heating — Some Details

Heat-flow calculations are extremely simple, notwithstanding most of the literature available. To leave the serious owner-builder with only a suggested insulation method without the actual specifications is insulting. On the other hand, presentation of insulation test data as commonly done in technical handbooks guarantees that sophisticated insulation design will remain a specialty of the heating engineer. I have simplified the presentation while retaining the accuracy of the information by compiling four simple tables that relate specifically to residential construction under average winter temperature conditions. These conditions are: 1) average temperatures inside the building section of 50° F, and 2) temperature difference between inside and outside of 30° F. That is, the data is exact for an inside temperature of 65° F and an outside temperature of 35° F — not far from the average conditions for the northern United States for the months of December, January, February, and March. Other temperatures would change the calculations by less than 10 percent.

Heat Flow

Heat is a form of energy. In reality, it represents the energy of motion of the atoms and molecules that make up a material. This energy always naturally flows from a warmer region to a cooler region. (The air conditioner is an active and therefore artificial machine, which reverses the natural flow by the expenditure of even more energy.) Energy is transferred from a warmer to a cooler region by three processes: conduction, convection, and radiation.

Conduction is the process by which the frying pan handle gets hot. The adjacent molecules of a material play a game of "pass it on" with the energy. A "conductor" is a material in which this heat transfer is rapid. An "insulator" is a material in which the process is slow. Of course, there is a whole spectrum of conductivities; it's all relative.

Metals are generally excellent conductors; air is about the poorest. A perfect vacuum has

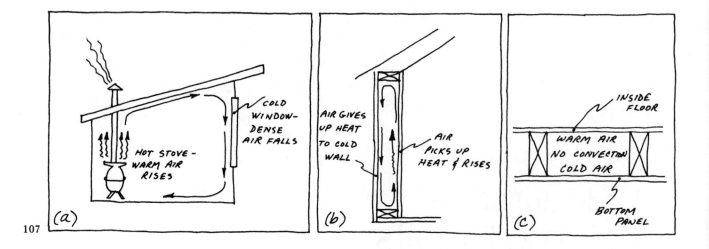

absolutely zero conductivity because it has no molecules to pass the heat. Among common building materials, dry wood is fairly low in conductivity while earth and masonry are fairly high.

Convection is the carrying of heat energy from one place to another by moving molecules. This time the molecules pick up energy in a warm region, carry it with them over a distance, and then deposit it in a widely separated place. The processes of picking up and dropping the heat are both conduction, but the carrying is convection. Air is a poor conductor but a good convector. Convection in air is caused by the fact that warm air is less dense and therefore rises, just like a hot-air balloon. The degree to which convection occurs depends upon the temperature differences in the air, the vertical distance of movement, and the resistance or friction presented to such movement. Illustration 107 shows three different examples of convection.

In A the temperature difference between air in contact with the stove and air in contact with the window is great, the vertical excursion is great, and the resistance to flow of the large room is small. Therefore, the convection is strong, the air making a round trip

in just a few minutes. Cigarette smoke released anywhere along a surface of the room will be observed to flow (convect) rapidly in the direction shown. In B, the temperature difference between inside and outside walls is great and the vertical distance great, but the frictional resistance to flow is also great. The air moving up along the inside wall is so close to the air moving down along the outside wall that they constantly bump into each other, slowing down the circular movement. In fact, the thinner the air space in the wall the less heat is convected! In C, the air just below the floor is warmed by conduction from the room above; the air at the bottom of the joist space is cooled by the cold panel. But this is a stable situation. That is, there is no physical reason to move, the hot and cold air already being where they want to be. Therefore, there is *no convection*. As we will see in the tables that follow, the thicker this air space, the less its conductivity.

Almost without exception, commercial insulating materials merely stop air convection by trapping it (closed cell foams) or slow it down greatly by friction (fibers, woods, open cell foams). The ultimate goal is represented by the conductivity of truly *dead air*.

Radiation is the means by which light is

transferred through absolutely nothing (space). Radiation from the sun travels 93 million miles through space losing nothing, about 50 miles through the atmosphere losing about half of its energy, and finally is stopped by your skin where it is changed into sensible heat, or heat that we feel. Light is merely the small percentage of radiation wavelengths to which our eyes are sensitive. A stove or radiator gives off radiation at longer wavelengths than the visible. In fact, all objects are radiating to each other continuously. The amount of radiation given off is proportional to the fourth power of the objects' temperatures. All we need to know is that more heat is transferred from hot to cold surfaces than from cold to hot, or *net heat is radiated from hot to cold.*

Some surfaces accept nearly all of the radiation falling on them. Others reject or reflect radiant energy. Most surfaces that readily accept radiant energy falling on them also give off or radiate their own energy very well. The percentage of energy accepted is usually close to the percentage radiated (not reflected) and is defined as the emissivity, *E*. The emissivity of black is high (more than 0.9, or 90 percent). Stoves are "blacked" to make them better radiators. Black roofs get hot in the sun because they accept the sun's energy, instead of reflecting it. Builder's foil, a highly polished aluminum foil on draft-paper backing, has an emissivity of 0.05 or less, meaning that 95 percent of all radiant energy falling on it is rejected or reflected back. When two surfaces face each other across an air space, the amount of heat transferred by radiation is determined by the emissivity of both surfaces, leading to an apparent emissivity for the space less than the emissivity of one surface alone. Table 10 lists emissivities of common building surfaces.

Table 10. *Emissivity of Various Surfaces*

Nature of Surface(s)	Emissivity, E
Ordinary boards, one surface	0.90
Ordinary boards, two facing surfaces	0.82
Galvanized steel, white or aluminum painted paper, one or two facing surfaces, Average	0.20
Builder's foil, highly reflective, one surface	0.05
Builder's foil, highly reflective, two facing surfaces	0.03

CALCULATING HEAT FLOW

We have already defined the measure of heat energy, the BTU. To refresh our memories: the BTU is the amount of heat required to raise the temperature of 1 pint of water by 1 F°.

The formula that describes heat flow by conduction is extremely simple.

(Equation 1)
$$H = \frac{A\Delta T}{R}$$

where

H = *total heat flow in BTU/hour*

A = *the area of building surface in sq.ft. through which heat is flowing*

ΔT = *the temperature difference in F° between inside and outside*

R = *the thermal resistance of the building section of area A*

The greater the area, the more heat flows; the greater the temperature difference, the more heat flows; and the greater the resistance to flow, the less heat is lost. Thermal resistance, *R*, is just the inverse of thermal conductance.

If we had to calculate the heat lost by a building using the exact laws of heat flow, we would indeed have to be heating engineers! Fortunately, all of the processes of heat flow (conduction, convection, and radiation) can be *assumed to be conduction* to a high degree of accuracy merely by assigning to parts of the building surface values of thermal resistance, *R*, which, when plugged into the conduction formula, give the proper answer. And this is exactly what the heating engineer does!

The typical building section (floor, wall, or roof) consists thermally of three aspects.

Table 11. *Thermal Resistance of Surfaces, R*

Nature of Single Surface	Direction of Heat	Emissivity, E		
		0.05	0.20	0.90
Inside wall	Horizontal	1.70	1.35	0.68
Inside ceiling	Upward	1.32	1.10	0.61
Inside floor	Downward	4.55	2.70	0.92
Outside surface, 15 mph	Any	—	—	0.17
Outside surface, 7-1/2 mph	Any	—	—	0.25

Surfaces, Table 11: Each section has two surfaces, one inside the building and one outside. The effective thermal resistance, *R,* of the inside surface is due to the thin boundary layer of dead air created along the surface by friction and the emissivity of that surface. The outside surface is subject to wind, which makes the boundary layer negligibly small, and the outside of a house typically has a medium value of emissivity, being neither black nor reflective as foil.

Air Spaces, Table 12: Heat is transferred across air spaces within building sections by all three processes. The relative contribution to the effective thermal resistance, *R,* of each is a function of thickness of the air space (conduction and convection); orientation of the air space relative to direction of heat flow (convection); and emissivity, *E,* of the combination of surfaces (radiation).

Building and Insulating Materials, Table 13: Heat is transferred through building materials almost entirely by conduction; Table 13 lists all of the building and insulation materials I could conceive of your using. Some of the items are given as a total thermal resistance, e.g., carpet. Most, however, are given as thermal resistance per inch of thickness to avoid a table about thirty-seven pages long! To get the total thermal resistance, simply multiply resistance per inch by inches of thickness. *Example:* For a 6-inch fiberglass batt: $R = 3.33 \times 6 = 20$.

Table 12. *Thermal Resistance of Air Spaces @ $T_{ave} = 50°$ F, $\Delta T = 30$ F°*

Air Space Location	Direction of Heat Flow	Air Space (in.)	Emissivity, E			
			0.03	0.05	0.20	0.82
Horizontal ex. ceiling	Upward	3/4	1.72	1.67	1.37	0.78
		1-1/2	1.82	1.76	1.43	0.80
		4	2.14	2.06	1.62	0.85
Horizontal ex. floor	Downward	3/4	3.80	3.55	2.39	1.02
		1-1/2	6.41	5.74	3.21	1.14
		4	10.70	8.94	4.02	1.23
Vertical ex. wall	Horizontal	3/4	2.95	2.80	2.04	0.96
		1-1/2	2.90	2.76	2.02	0.96
		4	2.74	2.63	1.94	0.94
Pitched 45° ex. roof in winter	Upward	3/4	2.02	1.95	1.54	0.83
		1-1/2	2.09	2.01	1.58	0.84
		4	2.32	2.22	1.71	0.88
Pitched 45° ex. roof in summer	Downward	3/4	3.47	3.27	2.27	1.01
		1-1/2	3.50	3.30	2.29	1.01
		4	3.61	3.39	2.33	1.02

Table 13. *Thermal Resistances of Building and Insulating Materials @ T_{ave} = 50° F, $\Delta T = 30°$ F*

Material	R (per inch)	R (total)
Softwoods	1.25	—
Hardwoods	0.91	—
Plywood	1.25	—
Poly vapor barrier	—	0.00
Sheetrock	0.90	—
Stone or concrete	0.08	—
Plaster or bricks	0.20	—
8-inch concrete block	—	1.11
1/8-inch cork tile	—	0.28
Cork board, average	3.50	—
Carpet with rubber pad	—	1.20
Carpet with fiber pad	—	2.1
Asphalt roll roofing	—	0.15
Asphalt selvage	—	0.30
Asphalt shingles	—	0.44
Standard exterior door	—	2.00
Insulating glass, 1/4-inch space	—	1.50
Insulating glass, 1-inch space	—	2.00
Fiberglass blanket or batt	3.33	—
Fiberglass or mineral loose fill	3.75	—
Polyurethane, 2 lb./cu.ft.	6.25–7.50	—
Polystyrene, blown	4.06–5.50	—
Polystyrene, bead board	3.70–4.25	—
Urea formaldehyde	4.50	—

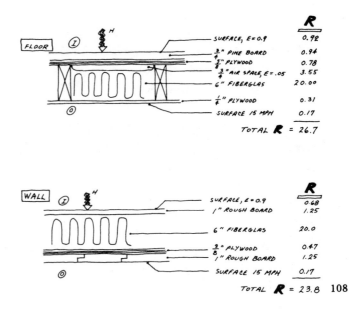

FINDING THE TOTAL THERMAL RESISTANCE

To find the total thermal resistance of a building section, simply: draw a picture of the construction so that you can identify all of the different heat-flow aspects; list each aspect and its thermal resistance; add all of the thermal resistances in the column to find the total thermal resistance (see Illustration 108).

INFILTRATION

Heat is lost by one final process, infiltration. As the term implies, warm inside air leaves the building as cold air rushes in to replace it.

Heat must be supplied to raise the incoming (infiltrating) air to room temperature:

(Equation 2) $H_{inf} = 0.0182Q\Delta T$

where

H = *heat required in BTU/hour*

Q = *rate of flow of air in cu.ft./hour*

ΔT = *room temperature minus outside temperature, $F°$*

Old houses may lose more than half of their heat to infiltration around old, ill-fitting windows and doors. Sometimes a large percentage even comes right up between those lovely old pumpkin pine floors! Even new houses, poorly constructed and containing a large number of openable windows, can lose as much as one-third of their heat by infiltration. A young couple just told me of their sad experience buying a brand-new house. It seems that the appealing idea of the house was a genuine antique hand-hewn barn frame sheathed with plywood and insulated with rigid polystyrene boards. It all sounded like high technology, but the missing element was the pride of the workmen, who made no attempt to seal the gaps between the plywood and the rough hand-hewn beams. The resulting infiltration required over a hundred tubes of caulking compound to seal the inside. The caulk can still be seen glistening between the boards and beams.

Any attempt to measure the length and width of all the cracks in an old building would not be worth the effort. I assume that you will have pride enough to make sure that there is essentially no infiltration through the building sheathing. If you take my advice and build your own fixed windows and refrigerator-type doors, you will eliminate those sources as well. My house is so tight that closing the back door pops the front door open against its spring by at least a foot — just like in a Volkswagen!

That doesn't mean that I'm against all ventilation. I'm only against ventilation I can't control! My method is to design a house with zero infiltration and then provide whatever controlled ventilation is needed. The amount of ventilation required is a function of the number of people in the house, their levels of activity, whether they take baths, the number of oxygen-producing plants, the oxygen requirement of the heating equipment, and what's cooking for dinner, to name a few. The amount of ventilation required in a well-insulated house is a lot less than generally believed.

I shoot for a figure somewhere between two extremes:

Smelly People Method (A). The building codes usually suggest a minimum air exchange rate of 10 cubic feet per person per minute. For a family of four, therefore

$$Q = 4 \times \frac{10 \; cu.ft.}{min.} \times \frac{60 \; min.}{hr.} = 2,400 \; \frac{cu.ft.}{hr.}$$

This would be an upper limit for my family because the occupancy rate of the building is only about 50 percent, we take a lot of baths, and we like house plants.

Combustion Method (B). If we assume that all of the air leaving the house is that consumed by the furnace or wood-burning stove in combustion, then the heat lost to infiltration or ventilation can be expressed as a percentage of the heat gained by combustion. If we assume a stove efficiency of 50 percent, and ΔT of 40 F°, then for every BTU of heat gained from the stove, we must supply .015 BTU to warm the makeup air. That is, after we figure the total BTU requirement of the house for the entire heating system, we must add 1.5 percent for infiltration.

In my own house, method A says that I should increase my annual heat load by 10 percent to account for a proper amount of ventilation. Method B says that my stoves require an increase of the annual heat load of about 1 percent. Both of these methods therefore show that required infiltration represents an insignificant heat load. I generally use 10 percent of the annual heat load, provided the kitchen stove and bathroom are mechanically ventilated.

CALCULATING MAXIMUM AND ANNUAL HEAT LOADS

The *maximum heat load* of a house is the number of BTUs per hour required to maintain the maximum ΔT of the year. In extreme northern states this is sometimes taken as $\Delta T = 90°$ F. In my locality, the average lowest annual temperature is –10° F. I'm willing to let the temperature of the house drop to 50° F that night, giving me a design value of $\Delta T = 60°$ F. Check your local building code. If there is no required value, the choice is yours.

To find the maximum heat load we simply use the conduction formula, Equation 1, for all the surfaces of the house (floor, walls, roof, windows, and doors) and add the results.

$$H_{max} = SUM\left(\frac{A\Delta T}{R}\right)$$

where $H = BTU/hour$

   $A = $ area of each surface in sq.ft.

   $R = $ thermal resistance of each surface

   $\Delta T = $ design maximum ΔT, $F°$

For an example we'll use the small house shown in Illustration 109, which has the building section insulation values calculated above.

To calculate the annual heat load we need

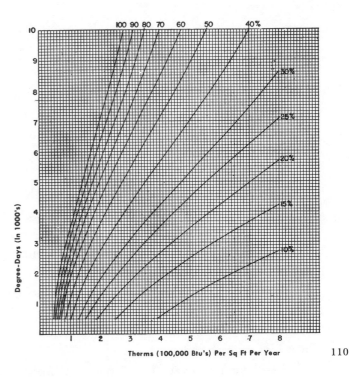

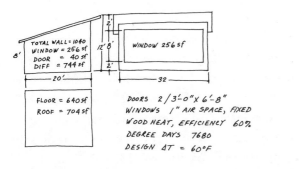

TOTAL WALL = 1040
WINDOW = 256 sf
DOOR = 40 sf
DIFF = 744 sf

FLOOR = 640 sf
ROOF = 704 sf

WINDOW 256 sf

DOORS 2 / 3'-0" X 6'-8"
WINDOWS 1" AIR SPACE, FIXED
WOOD HEAT, EFFICIENCY 60%
DEGREE DAYS 7680
DESIGN ΔT = 60°F

SECTION	R	A (sf)	ΔT	$H_{max} = \frac{A\Delta T}{R}$	$H_{annual} = \frac{THERMS \times A}{R}$
FLOOR	26.7	640	60	1440 BTU/HR	86.3
ROOF	29.3	704	60	1440	86.5
WALL	23.8	744	60	1876	112.5
WINDOW	2.0	256	60	7680	460.8
DOOR	5.5	40	60	440	26.2
INFILTRATION				644	40
				13,500 BTU/HR	812 THERMS/YR

109   SMALL HOUSE HEAT LOAD

two additional pieces of information: the number of degree days for the site and the efficiency of the heating device. I suggest you use efficiencies of: gas furnace, 80%; oil furnace, 70%; coal furnace, 60%; antique wood stove, 40%; modern airtight wood stove, 60%; open fireplace or stove, 15%; electric baseboard, 100%.

Procedure:

(1) Enter Illustration 110 with degree days and efficiency. Locate degree days on the verti-

cal scale and come across until you strike your efficiency curve. From that point, drop straight down and read out therms/sq.ft./year. (1 therm = 100,000 BTU).

(2) Multiply this number by A/R for each building section.

(3) Add figures for entire building to get total therms per year for the building.

(4) Multiply therms per year by the conversion factor in Table 14 to find fuel amounts.

Using the same building, let's calculate the annual fuel requirement for Portland, Maine (7,680 degree days), using air-dried hardwood and an airtight wood stove (efficiency 60 percent). Entering Illustration 110 with 7,680 degree days, we come across to the right until we strike the 60 percent curve. Dropping straight down we find 3.6 therms per year per sq.ft. Multiplying 3.6 by A/R for each building section, we get the values shown. The total, including 5 percent infiltration, is 812 therms per year.

Multiplying this figure by the factors in

Table 14, we get the following fuel requirements: 2.7 cords air-dried hardwood or 5.4 cords air-dried softwood.

Repeating the whole procedure using the efficiencies for coal, oil, and electric heat, we find the same house would require 3-1/3 tons anthracite coal, 494 gallons No. 2 fuel oil, or 14,300 kilowatt-hours electricity.

In the example we have assumed heat loss through the windows but no solar heat gain. The definition of the degree day is predicated upon a uniform distribution of windows around the house. Part of the difference between 72° F and the 65° F base is assumed to be due to solar heat gain. Even so, however, about one-half of the total heat load of the house is nighttime heat loss through the windows.

We will next analyze the *passive solar house*, a house designed to utilize the sun to the maximum extent without resorting to expensive active collectors and storage. We will find that a house of the same area can be operated in a manner to capitalize on the sun's heat to a remarkable degree.

Table 14. *Conversion of Therms to Fuel Quantity* (Multiply therms by factor to obtain quantities of fuels)

Factor	Fuel Type
10.1	Lb. coal @ 10,000 BTU/lb.
8.3	12,000 BTU/lb.
7.1	14,000 BTU/lb.
0.71	Gal. No. 2 fuel oil
100	Cu. ft. gas @ 1,000 BTU/cu. ft.
125	800
0.0033	Cords hardwood @ 20% moisture (air dried)
0.0030	@ 12% (inside dried)
0.0063	Cords softwood @ 20% moisture
0.0058	Cords softwood @ 12% moisture
29.3	Kwh electric heat

same floor area is computed for twenty-two different permutations, ranging from two-thirds south window wall and no night window cover to a house with 100 percent glass south wall, rigid insulating shutters, and a heat-storing concrete slab floor. These houses represent varying degrees of the passive solar house. To achieve a higher percentage of solar heating would require an *active solar house*, one in which heat is actively (mechanically) removed from roof collectors and stored for later use.

The Passive Solar House

In the previous example we found that more than half of the annual heat load was lost through the double-glazed windows. However, we had taken little advantage of either 1) heat gained as solar radiation through south-facing windows, or 2) the possibility of decreasing nighttime heat loss by further insulating the windows.

We here present an extension of the analysis in which the annual heat load for a house of the

ACTIVE VS. PASSIVE SOLAR

As we will see in the following analysis, there is great potential for utilizing the sun's energy to balance nighttime heat losses. However, as usual, the industries capitalizing upon this potential have unfortunately distorted the picture for the public. The picture sometimes painted is a solar collector, bought from a local supplier, which, when placed upon a roof inclined 60° to the south, will supply more than 80 percent of the house heating load. An implicit assumption — rarely denied — is that continuing research by American technologists will increase the efficiency of these devices until, ultimately, a small collector placed outside the house or perhaps in a window will "magnify" or "compress" the heat enough to provide all of the heat.

I hope not to put a damper on people's enthusiasm for solar heat. The enthusiasm is needed as we enter the post-industrial age. However, we stand a better chance of long-term energy self-sufficiency if we make design decisions based upon hard facts.

Fact 1: The amount of solar energy received from the sky is fixed by solar geometry and the weather. This energy is listed in tables and charts as BTU/sq.ft. per minute, hour, or day. No invention will ever increase the radiation density received. That is, if the chart gives the measured value as 200 BTU/sq.ft. per hour, then a collector 1 sq.ft. in area will be required to gather 200 BTU/hour — period!

Fact 2: Any solar collector constructed of ordinary materials (glass, plastic, metal, wood) has an upper limit of efficiency (% collected/% received) determined by collection temperature versus outside temperature, transmittance of the transparent cover, etc. Collectors of reasonable cost do not presently exceed the efficiency of ordinary windows having the same numbers of cover glasses. All light received through a window is collected directly as heat at the minimum usable temperature.

Fact 3: If the collector is to be used only for winter heating and not summer air conditioning, during the time when the ground is covered with snow a south wall will receive more total radiation than a roof collector inclined at 60°. Therefore, for the most northern states of the United States, the south wall is often the ideal collecting surface.

Fact 4: Disadvantages of tilted roof collectors include:

(1) the possibility of damage by falling objects;

(2) the difficulty of servicing;

(3) added cost for a part of the building serving no other function;

(4) a compromise in building design. The ideal tilt lies between 45° and 90°, depending upon location and whether it is to be used all year or just in the winter. Therefore it is either a roof of greater than normal pitch or a vertical wall competing directly with south-facing windows with their obvious and spectacular psychological benefits.

(5) collection of heat requires ductwork, motors, and elaborate controls. There is potential for noise and leakage. The complexity is far greater than the average furnace and will probably require greater maintenance.

(6) the initial cost of an active solar system in the northern United States capable of supplying 50 percent of the heat load will never fall below $5 to $10 per sq.ft. of living area, depending upon whether you do it yourself or have it installed. In contrast, the passive solar house will cost an extra $1 to $2 per sq.ft. of living area.

All of the above facts lead to our consideration of the passive solar house, a house designed in principle to:

(1) admit the maximum amount of winter light (heat) through a south window wall;

(2) lose minimal heat through all other surfaces;

(3) trap gained heat by nighttime insulating devices over the window wall;

(4) prevent excessive summer heat gain by a south roof overhang.

The dimensions of the house are shown in Illustration 111 along with the calculated section thermal resistance values. Note that this house has the same floor area as the house in the previous example, allowing comparison of its thermal performance to that of a conventional house. The house is to be considered as a mod-

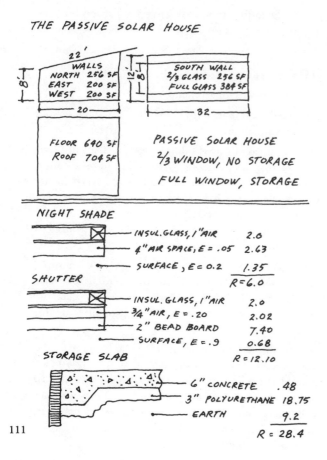

THE PASSIVE SOLAR HOUSE

WALLS
NORTH 256 SF
EAST 200 SF
WEST 200 SF

SOUTH WALL
2/3 GLASS 256 SF
FULL GLASS 384 SF

FLOOR 640 SF
ROOF 704 SF

PASSIVE SOLAR HOUSE
2/3 WINDOW, NO STORAGE
FULL WINDOW, STORAGE

NIGHT SHADE

INSUL. GLASS, 1"AIR   2.0
4" AIR SPACE, E=.05   2.63
SURFACE, E=0.2   1.35
                         R=6.0

SHUTTER

INSUL. GLASS, 1"AIR   2.0
3/4"AIR, E=.20   2.02
2" BEAD BOARD   7.40
SURFACE, E=.9   0.68
                        R=12.10

STORAGE SLAB

6" CONCRETE   .48
3" POLYURETHANE   18.75
EARTH   9.2
                       R=28.4

111

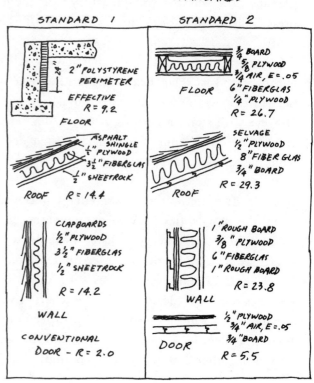

INSULATION STANDARDS

STANDARD 1

2" POLYSTYRENE PERIMETER
EFFECTIVE R=9.2
FLOOR

ASPHALT SHINGLE
1/2" PLYWOOD
3 1/2" FIBERGLAS
1/2" SHEETROCK
ROOF R=14.4

CLAPBOARDS
1/2" PLYWOOD
3 1/2" FIBERGLAS
1/2" SHEETROCK
R=14.2
WALL

CONVENTIONAL
DOOR - R=2.0

STANDARD 2

3/4" BOARD
5/8" PLYWOOD
3/4" AIR, E=.05
6" FIBERGLAS
1/4" PLYWOOD
R=26.7
FLOOR

SELVAGE
1/2" PLYWOOD
8" FIBER GLAS
3/4" BOARD
R=29.3
ROOF

1" ROUGH BOARD
3/8" PLYWOOD
6" FIBERGLAS
1" ROUGH BOARD
R=23.8
WALL

1/2" PLYWOOD
3/4" AIR, E=.05
3/4" BOARD
R=5.5
DOOR

ule. To determine the heat loads for small (640 sq.ft.), medium (1,280 sq.ft.), and large (1,920 sq.ft.) houses, simply multiply the module heat load by 1, 2, or 3.

Three basic houses are considered:

House I is insulated to the present housing industry standard, described in Illustration 111 as Standard 1. All of the windows are concentrated in the south wall, which has a resulting window ratio of two-thirds. The ratio two-thirds is

Table 15. *Portland, Maine, Degree Days Saved, 2/3 South Window Wall*

Month	1. Degree Days @65° F	2. Degree Days @55° F	3. Hours of Sun	4. % Sun Heat	5. ΔT Adj.	6. ΔDD @ΔT	7. DD Saved @65° F	8. DD Saved @55° F
October	515	215	9	28	-2.3	-69	75	0
November	825	525	8	25	-2.1	-63	139	69
December	1,237	937	7-1/3	23	-1.9	-57	228	159
January	1,373	1,073	8	25	-2.1	-63	280	205
February	1,218	918	9	28	-2.3	-69	272	188
March	1,039	739	10	31	-2.5	-75	247	154
April	693	393	11-1/3	35	-2.6	-78	164	60
Total DD Saved							1,405	835
Aux DD							6,275	3,737
Fuel Factor (therms)							X2.9	X1.7

actually greater than optimal without thermal storage capability. Any larger glass ratio than one-third has been found to result in overheating even on average days unless thermal mass (masonry or water) has been included in the building's interior.

House II is the same two-thirds south window wall house but insulated to Standard 2, which is minimal for the northern United States.

House III has a full glass window wall and is built upon a 6-inch concrete slab insulated from below. The thermal capacity of the slab stores excess daytime radiation without venting and gives it back over the average 16-hour January night, dropping in temperature only 10° F.

The annual heat load of each of the three houses is calculated with three different night-time window treatments: 1) no cover, 2) a

HOURLY INSOLATION ON CLOUDLESS DAYS

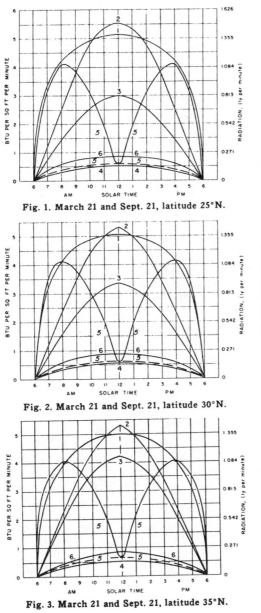

Fig. 1. March 21 and Sept. 21, latitude 25°N.

Fig. 2. March 21 and Sept. 21, latitude 30°N.

Fig. 3. March 21 and Sept. 21, latitude 35°N.

In the illustrations, the numbers on the curves refer to the surface on which energy is received as follows:

(1) Normal incidence.
(2) Horizontal surface.
(3) Vertical surface facing South.
(4) Vertical surface facing North.
(5) Vertical surface facing East during the morning and West during the afternoon. Dashed lines are for East and West walls during afternoon and morning respectively.
(6) Diffuse on horizontal surface.

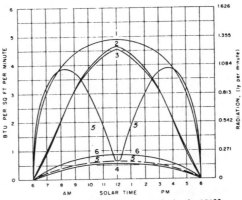

Fig. 4. March 21 and Sept. 21, latitude 40°N.

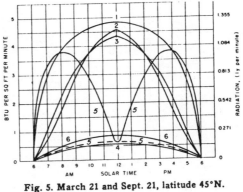

Fig. 5. March 21 and Sept. 21, latitude 45°N.

112A

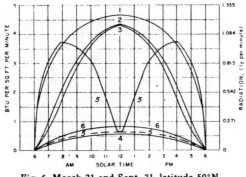

Fig. 6. March 21 and Sept. 21, latitude 50°N.

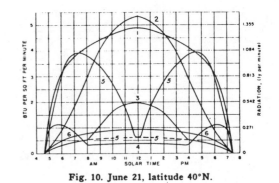

Fig. 10. June 21, latitude 40°N.

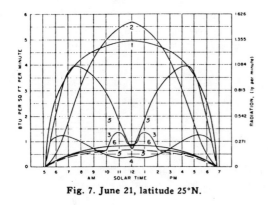

Fig. 7. June 21, latitude 25°N.

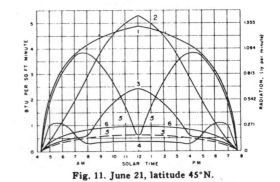

Fig. 11. June 21, latitude 45°N.

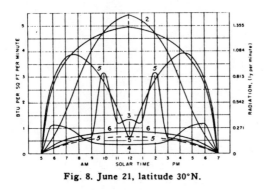

Fig. 8. June 21, latitude 30°N.

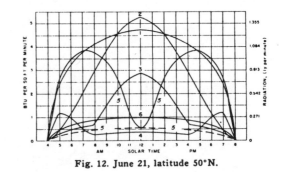

Fig. 12. June 21, latitude 50°N.

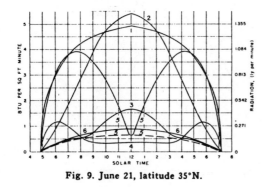

Fig. 9. June 21, latitude 35°N.

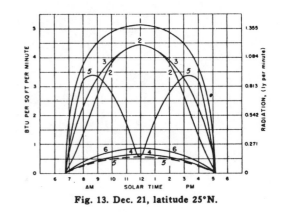

Fig. 13. Dec. 21, latitude 25°N.

112B

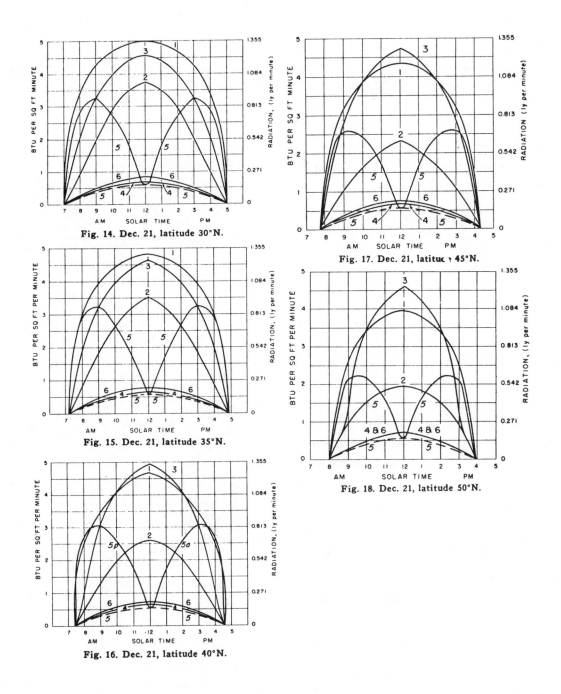

Fig. 14. Dec. 21, latitude 30°N.

Fig. 15. Dec. 21, latitude 35°N.

Fig. 16. Dec. 21, latitude 40°N.

Fig. 17. Dec. 21, latitude 45°N.

Fig. 18. Dec. 21, latitude 50°N.

112C

reflective foil shade, and 3) a rigid shutter. In addition, all of the above permutations are calculated assuming two different mean inside temperatures: 72° F (65° DD base) and 62° F (55° DD base).

Calculations for Houses I and II: The basic assumption that allows us to calculate the annual heat load of Houses I and II is that a predictable percentage of days will have enough solar radiation (the critical intensity) to supply the entire heat load during the daylight hours. The percentage of time thus solar-heated results in an apparent reduction in degree days. The calculation is done on a monthly basis for the seven-month heating season from October to April.

Table 15 (page 174) shows tabulated calculations. Column 1 lists the published values of monthly degree days for Portland, Maine. Column 2 lists monthly degree days based upon a 55° F base. The latter values were obtained simply by subtracting 300 DD for each month in Column 1.

Next, the critical intensity of solar radiation was computed by equating solar heat gain through the windows to the hourly heat loss from the entire house at a ΔT of 40° F (representing the difference between average inside temperature, 72° F, and average daytime outside temperature, 32° F, for the month of January). The critical intensity was found to be 78 BTU/sq.ft. per hour for House I and 52 BTU/sq.ft. per hour for House II.

In Illustration 112A–C (pp. 175–7), number of hours of above-critical solar radiation on clear days for December 21, September 21, and March 21 is found. The number of hours for other months can be interpolated using comparative times of sunrise and sunset. Clear days have excessive radiation that must be vented from the house at midday. Partly cloudy and bright

cloudy days also can have sufficient radiation. Illustration 113 shows the percentage of days versus total daily radiation on a south wall in January actually measured in Boston, Massachusetts. Illustration 114 shows that Portland and Boston have equivalent solar weather. Using an average critical intensity of 60 BTU/sq.ft. per hour, we find that three-quarters of all January days have sufficient radiation to balance heat loss. That figure is confirmed by the authors, who live in houses similar to House II.

Column 3 lists hours of solar heat on clear days. Column 4 is the monthly percentage of hours of solar heat, and is obtained by multiplying Column 3 by 3/4 and dividing by 24 hours.

Columns 5 and 6 are a technical adjustment that reflect the fact that the daylight hours contribute fewer degree days percentage-wise (they are warmer). Therefore we must put back some of the degree days thought to have been saved.

Column 7 is degree days saved by solar heat and is computed as Column 1 × Column 4, plus Column 6. Column 8 is degree days saved on the 55° base and is calculated as Column 2 × Column 4, plus Column 6.

Finally, the total degree days saved is tallied and subtracted from the published annual degree days to obtain the fuel degree days, or the degree days that must be supplied by fuel. The fuel degree days are used with Illustration 110 to obtain the fuel factors 2.9 and 1.7 therms. Having these factors it is a simple matter to compute the fuel requirements shown in Tables 16 (House I) and 17 (House II).

Calculations for House III (Slab Heat Capacity): The same general degree-day-saving approach is used in calculating the annual heat load for House III. However, we now must find an

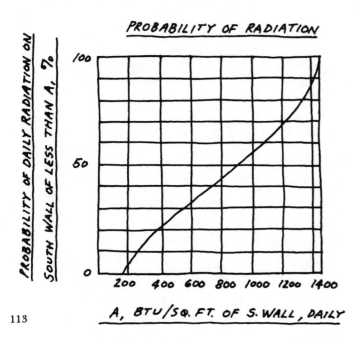

PROBABILITY OF RADIATION

PROBABILITY OF DAILY RADIATION ON SOUTH WALL OF LESS THAN A, %

A, BTU/SQ. FT. OF S. WALL, DAILY

113

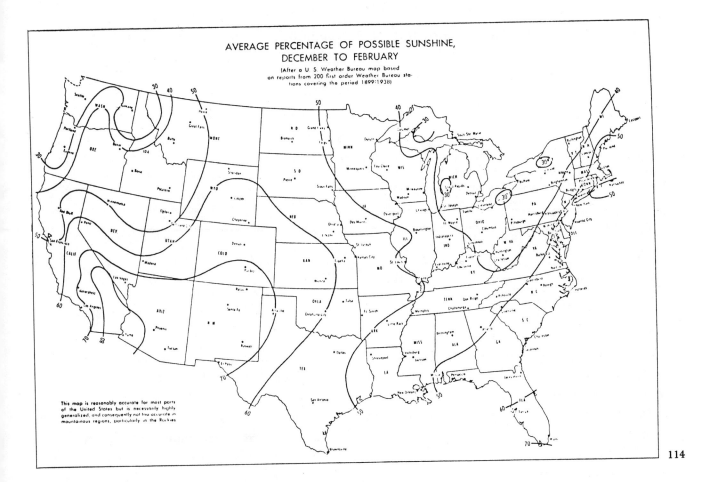

AVERAGE PERCENTAGE OF POSSIBLE SUNSHINE, DECEMBER TO FEBRUARY

(After a U.S. Weather Bureau map based on reports from 200 first order Weather Bureau stations covering the period 1899:1938)

This map is reasonably accurate for most parts of the United States but is necessarily highly generalized, and consequently not too accurate in mountainous regions, particularly in the Rockies.

114

Table 16. *Annual Heat Load, 640 sq.ft. Module, 2/3 South Window, Insulation Standard 1*

Section	A(SF)	R	A/R	R_{ave}	No Sun 3.6	Insulating Glass 2.9	Astrolon® Shade 2.9	Shutter 2.9	No Sun 2.1	Insulating Glass 1.7	Astrolon® Shade 1.7	Shutter 1.7
						65° F Base				55° F Base		
Floor	640	9.2	69.6	—	250	202	202	202	146	118	118	118
Roof	704	14.4	48.9	—	176	142	142	142	103	83	83	83
Wall	744	14.2	55.2	—	207	160	160	160	116	94	94	94
Door	40	2.0	20.0	—	72	58	58	58	42	34	34	34
Insulating Glass	256	2.0	128	7.4	461	371	—	—	269	218	—	—
Night Shade	256	5.5	46.5	9.9	—	—	135	—	—	—	79	—
Shutters	256	12.0	21.3	10.1	—	—	—	62	—	—	—	36
Infiltration					40	40	40	40	24	24	24	24
Annual Heat 10^5 BTU (Sum of Sections)					1,266	973	737	664	700	571	432	389
No. 2 Oil Gallons @70% Efficiency					900	592	448	404	427	348	263	237
Mwh Electricity @ 100% Efficiency					22	17	13	12	12	10	8	7
Cords Hardwood @60% Efficiency					4.2	3.2	2.4	2.2	2.3	1.9	1.4	1.3

Table 17. *Annual Heat Load, 640 sq.ft. Module, 2/3 South Window, Insulation Standard 2*

Section	A(SF)	R	A/R	R_{ave}	65° F Base				55° F Base			
					No Sun 3.6	Insulating Glass 2.9	Astrolon® Shade 2.9	Shutter 2.9	No Sun 2.1	Insulating Glass 1.7	Astrolon® Shade 1.7	Shutter 1.7
Floor	640	26.7	24.0	—	86	70	70	70	50	41	41	41
Roof	704	29.3	24.0	—	86	70	70	70	50	41	41	41
Wall	744	23.8	31.3	—	112	91	91	91	66	53	53	53
Door	40	5.5	7.3	—	26	21	21	21	15	12	12	12
Insulating Glass	256	2.0	128	11.1	461	371	—	—	269	218	—	—
Night Shade	256	5.5	46.5	17.9	—	—	135	—	—	—	79	—
Shutters	256	12.0	21.3	22.1	—	—	—	62	—	—	—	36
Infiltration					40	40	40	40	24	24	24	24
Annual Heat 10^5 BTU (Sum of Sections)					811	663	387	314	474	389	250	207
No. 2 Oil Gallons @70% Efficiency					495	404	235	191	289	237	152	126
Mwh Electricity @ 100% Efficiency					14	12	7	6	8	7	4	4
Cords Hardwood @60% Efficiency					2.7	2.2	1.3	1.0	1.6	1.3	0.8	0.7

additional percentage of days in which the stored radiation provides the entire daily heat load via slab storage.

The average January overnight (16-hour) heat load is:

$$H = \frac{A\Delta T}{R_{ave}} \times 16 \ hr.$$

where

A = *house surface area*

ΔT = *ave. inside T minus ave. outside T*

= *60° F minus 10° F = 50° F*

R_{ave} = *effective thermal resistance of entire house (15.8 using shades of R = 5.5)*

H = *(2,384 × 50 × 16)/15.8*
= *120,700 BTU*

The thermal storage capacity of concrete (stone, too) = 38 BTU/cu.ft. per F°. Allowing the slab to cool by 10 F° in 16 hours, we require a

$$Volume = \frac{120,700 \ BTU}{\dfrac{38 \ BTU}{cu.ft./F°} \times 10° \ F} = 320 \ cu.ft.$$

Since the area of the slab is 640 sq.ft., and *Volume = Area × Thickness*, the slab thickness must be 6 inches. Equivalent conditions would be a 3-inch slab and 20 F° temperature drop.

A conservative simplifying assumption used is that the percentage annual degree day saving is identical to the percentage January degree day saving. This is a reasonable assumption since both maximum heat load and south wall radiation occur within a month of each other (January 21 and December 21). The calculations for House III are tabulated in Table 18.

Line 1 shows the effective R values for the house as a whole during the 8-hour day/16-hour night with the various window covers.

Lines 2 and 3 are the heat loads for the 8-hour day and 16-hour night using the R values of Line 1.

Line 4 is the total 24-hour heat load, or the sum of Lines 2 and 3.

Line 5 computes the 24-hour critical radiation; i.e., the daily amount of radiation per sq.ft. of south window that balances the 24-hour heat load.

Line 6 computes the 8-hour critical radiation, or the daily radiation per sq.ft. of window that balances the 8-hour daytime heat load.

Table 18. *Portland, Maine, Degree Days Saved, Full South Window, Slab Storage*

Assumptions:
% Annual Saving = % Jan. Saving
DAY = 8 hr. @ΔT = (70 – 30) = 40° F
NIGHT = 16 hr. @ΔT = (60 – 15) = 45° F
DD_{65} = 7,680, DD_{55} = 4,572
Transmission = 72%. Glass = 90%.

	65° F Base			55° F Base		
	Insulating Glass	Astrolon® Shade	Shutter	Insulating Glass	Astrolon® Shade	Shutter
1. $R_{ave, day}/R_{ave, night}$	8.7/8.7	8.7/15.8	8.7/21.1	8.7/8.7	8.7/15.8	8.7/21.1
2. Heat load$_{day}$ (10^3 BTU)	87.7	87.7	87.7	65.8	65.8	65.8
3. Heat load$_{night}$ (10^3 BTU)	197.3	108.6	81.3	153.5	84.5	63.3
4. Heat load$_{total}$ (10^3 BTU)	285.0	196.3	169.0	219.3	150.3	129.1
5. Critical radiation$_{24 hr}$ (BTU/SF)	1,145	788	679	881	604	519
6. Critical radiation$_{8 hr}$ (BTU/SF)	352	352	352	264	264	264
7. Percent solar heat$_{24 hr}$	33	57	61	50	66	70
8. Percent solar heat$_{8 hr}$	48	24	20	37	21	17
9. Solar supplied degree days	3,760	4,990	5,220	2,830	3,340	3,470
10. Auxiliary heat degree days	3,920	2,690	2,460	1,740	1,230	1,100
11. Fuel factor (therm/sq.ft., yr.)	1.7	1.3	1.2	1.0	0.8	0.8

Line 7, the percentage of days having the 24-hour critical radiation, is found from Illustration 112.

Line 8 is found by subtracting Line 7 from the percentage of days having the lower 8-hour critical radiation, again found from Illustration 112.

Line 9 computes the degree days supplied by solar heat. Note that no credit is made beyond 8 hours for those days having radiation levels *between* 8 and 24 hours. This is an extremely conservative assumption in the case of oil or electric heat, but corresponds to the probable mode of operation with wood heat. If the owner judges there to be insufficient stored heat for the night, he will light a fire regardless of the fraction in storage.

Line 10 is the difference between published degree days and solar-supplied degree days — or auxilliary heat degree days. Illustration 110 is entered with auxilliary heat degree days to find the appropriate fuel factor in therms/sq.ft.

per year on Line 11. Table 19 presents the resulting annual heat loads for House III.

Conclusions

There are four main variables in the economic feasibility of solar heat. We must make a clear distinction here. When I worked at MIT, we used to tell government sponsors that, given enough money, anything is feasible. We put a man on the moon for about a trillion dollars. Whether it was worth it depends upon your perspective. The same is true of solar heat. You can now buy solar heat panels that, when placed upon a roof of proper pitch and orientation, will provide more than 50 percent of the annual heat load. But, if you're like me, you'll want to know if this is the cheapest way to heat your house. I

Table 19. *Annual Heat Load, 640 sq.ft. Module, Full South Window, One-Day Heat Storage Capacity*

Section	A(SF)	R	A/R	R$_{ave}$	65° F Base			55° F Base		
					Insulating Glass 1.7	Astrolon® Shade 1.3	Shutter 1.2	Insulating Glass 1.0	Astrolon® Shade 0.8	Shutter 0.8
Floor	640	26.7	24.0	—	41	31	29	24	19	19
Roof	704	29.3	24.0	—	41	31	29	24	19	19
Wall	616	23.8	25.9	—	44	34	31	26	21	21
Door	40	5.5	7.3	—	12	10	9	7	6	6
Insulating Glass	384	2.0	192	8.7	326	—	—	192	—	—
Night Shade	384	5.5	69.8	15.8	—	91	—	—	56	—
Shutters	384	12.0	32.0	21.1	—	—	38	—	—	26
Infiltration					25	20	17	15	12	10
Annual Heat 10^5 BTU (Sum of Sections)					489	217	153	288	133	101
No. 2 Oil Gallons @ 70% Efficiency					298	133	93	175	81	61
Mwh Electricity @ 100% Efficiency					8.6	3.8	2.7	8.4	3.9	3.0
Cords Hardwood @ 60% Efficiency					1.6	0.7	0.5	1.0	0.4	0.3

can say with certainty that it is not at this date in the zone defined as having minimum feasibility. And in the heavily wooded rural areas of the United States it probably will not be for fifty years because the competing fuel is wood.

This chapter has shown how to compute fuel savings and annual heat load, given the two climatic variables of annual degree days and hours of sunshine. The other two variables, however, are *really* variable — the extra annual cost of the solar aspect of the house (whether active roof collectors or merely south windows with reflective shades) and the cost of fuel saved. Analysis of all of these factors is the only true basis for the economic decision, and it will vary with time and area of the country.

The conclusions for Portland are clearly shown in Table 16, 17, and 19. The geometric shift of windows from a conventional random orientation to a concentration on the south wall results in a fuel saving of 15 to 20 percent. The nighttime use of simple reflective foil shades saves from 25 to 50 percent, depending upon the insulation standard of the rest of the house. However, the further percentage saving resulting from the use of shutters ($R = 12$) rather than shades ($R = 5.5$) is surprisingly small and perhaps not economically justified.

The difference between Houses I and II is only in insulation standard. The use of 6-inch wall and 8-inch ceiling insulation rather than 3-1/2-inch saves about 50 percent in fuel requirement. If the FHA must have standards, let them be $R = 20$ walls and $R = 30$ ceilings instead of concrete foundations and 7-foot ceiling heights.

Running a house at 62° F instead of 72° F results in a fuel saving of 40 percent, which would finance new longjohns and sweaters for everyone, including the dog.

House III exhibits a saving of 25 to 40 percent over House II by the addition of the storage slab. However, the insulated slab is expensive, the primary cost being the 3-inch rigid urethane insulation. Also, the slab must not be covered by more than a wall-to-wall carpet in order to be effective, which may pose an aesthetic problem to some people.

In general, remember that fuel cost savings are due to fuel *amounts* and not *percentages*.

One might save 40 percent of the fuel bill in going to House III from House II, but the fuel amounts for a 1,280 sq.ft. house using reflective shades are only 2.6 and 1.6 cords of wood per year — a difference of 1 cord of wood per year. I can get hardwood cut, split, and delivered for $45 per cord; in 4-foot-lengths from a pulpwood hauler for $30; and in my backyard for the cost of gas for my chainsaw and a good day's work.

(CW)

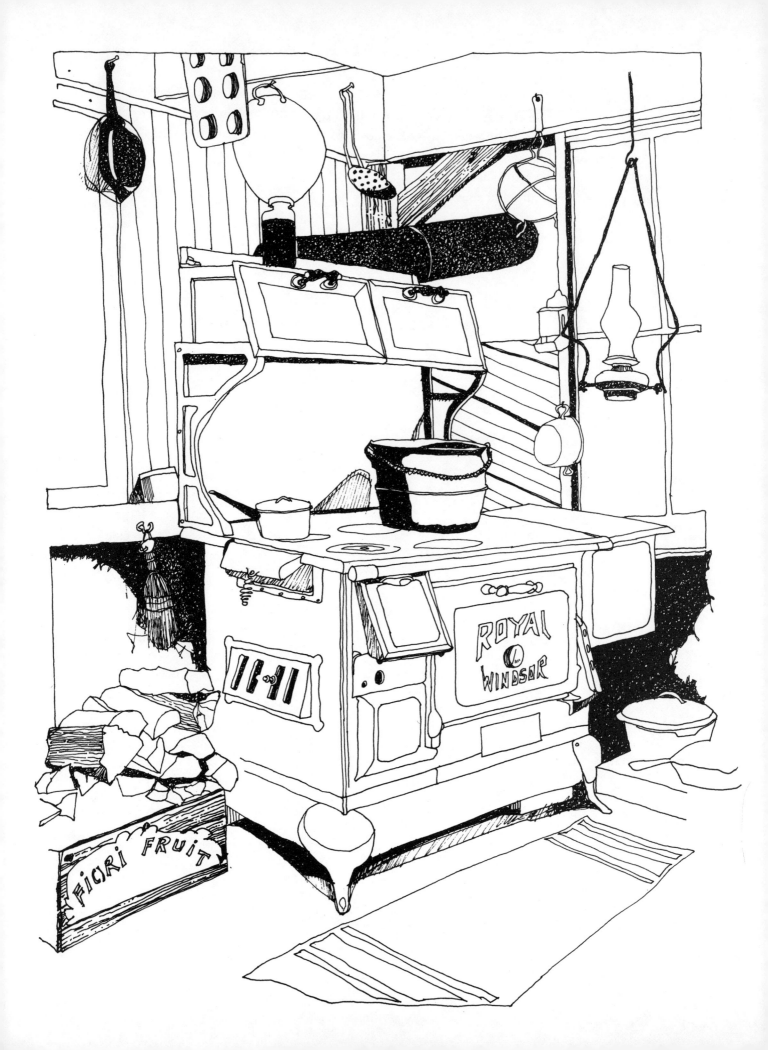

20

Heating with Wood

By the time November arrives in northern latitudes, wood burning becomes routine for those households that use the fuel as a primary or supplemental heat source. There is no measurement possible of the dreams, hopes, musings, and memories that take wing along the first bright flames and gentle tendrils of wood smoke, but the ratio would be impressive if the tally could be made — about a score of dreams and musings to every half-cord, I'd say.

In our shelter, Sam builds the fires that dreams are made of. It is his regular afternoon chore when he gets home from school. My part is to make certain there is wood ready — dry, split, and stacked in the barn and on the porch.

For a short time last November, the fires burned superbly. Sam was complimented on his firemaking skills, but there was a larger reason, a much larger one. It was a huge oak I had found in our woods earlier that autumn.

A decade or so before my discovery, the same woods had been cut over by woodsmen harvesting softwood — pine and spruce — for the local paper mill. The oak must have been wounded

when a large pine toppled, stripping the oak of most of its branches. The tree died slowly; when I found it, only the towering trunk remained — a silver-gray pillar, still standing straight among the evergreens. Where the bark had once been, there was the minute beginning of decay, but just beneath the gentle flaking my saw bit into red oak as hard as slate.

More than two weeks' wood for our two firepots was sawed and split from that gray giant. The wood was as seasoned as a section of antique furniture — dry, fine-grained, and as flinty as an ebony tabletop. Whenever I added a section of that oak to the fire, I felt a kind of ceremonial twinge, as if the bones of a hero were being cremated.

There was more than just a fanciful reason for my thoughts; the tree was extraordinary. As I had sawed it into sections for splitting, the fine steel of my bucksaw rasped on several sections of barbed wire — the remnants of some long-gone fence that had kept cattle from wandering more than a half-century before when the woodland had been cleared for pasture. My heroic oak had been a line tree — a natural presence on the boundary that separated the Orr farm from the Simpson place, or perhaps the Spinney land.

Those names outlived the farms. Today they identify local islands, roads, points, creeks, and marshes. The barns, homes, buildings, and fences those first families built are gone; only the cellar holes remain, buried among ancient lilacs and alders rushing to take over. The great oak I had found was one of the final survivors of those farming days.

I mused on those vanished families, the absent pastures, and the stumps of pine and spruce as I watched the flames curl so silently and steadily from the fine, gray logs. What would they think of the routine of our wood fires, I wondered. Could they comprehend the energy circle that had been closed in less than a century.

We light the fires in November, and every afternoon from then through April, because we use wood, the renewable resource, to supplement the other "renewable" that heats our home — the sun. Whenever it shines, even dimly, through the bank of double-pane windows on the south side, the sun gives our place enough warmth on even the coldest days to keep us cozy in our comfort zones.

Sam does his fire-building about a half-hour before those early December sunsets, and the firepots maintain the comforts that the sun has given. Only after we have gone to bed and the zero-degree outside becomes a midnight thief of the gathered warmth does the central oil-fired furnace kick on and bring us back into the petroleum age.

Often I am awake when that happens, and when I hear the rumble I yearn for the wind dynamo we have planned. Now abuilding, the wind machine will generate some 15 kilowatts of electricity whenever the wind blows 14 mph or better. During Maine winters, such winds are habitual. With the windmill on the point, feeding its power directly into resistance electric heaters in each room of the house, we'll be warmed by another renewable on seven of every ten of those winter days and nights. The louder the gales howl, the warmer we'll be.

I dream often in front of the wood fire about the day that windmill-yet-to-come will add its renewable energies to those of wood and sun. But I also ponder on the irony that it was just fifty years ago that the Simpsons, Orrs, and Spinneys saved and planned and sacrificed so they could put central heating into those farmhouses where the cooking was always done on wood cookstoves. And it was just ten years ago that the woodcutters walked over what had been farmland, harvesting a timber crop, heedless of the trees lost, the slash piled on the forest floor, or the line oaks stripped of their limbs.

Yet I do my dreaming at a time when Americans are realizing that they must seriously rethink such casual relationships with nature. For the first time in the two centuries since the nation was founded, the assumptions partner to central oil heat are being questioned; the routine harvesting of timber resources is no longer routine. The Orrs and Simpsons would be startled, yet I can show you the cellar holes of their homes that were here just two generations ago.

The change is profound. It has its roots in the national origins. Nowhere in the original papers of the nation are there references to man's relationships with nature. There are, in the Declaration of Independence, the Constitution, in the early correspondence of Franklin, Adams, Washington, and Jefferson, only designs and plans for man's relationship with man.

The narrowness is understandable. The discovery of this continent was a heady business. It gave the discoverers and developers a basis for believing that man had assumed command, that the great expanse of North America lay at his feet for him to exploit, manage, develop, and maintain. No sooner had that work begun than science and technology produced the tools that convinced man of his invincibility. Until the 1970s, nature has been considered merely the provider of the raw materials.

But now, burning wood, building wisely, accepting and saving the sun's light and warmth, we take small steps toward an accommodation with nature because we have become aware that many raw materials are nonrenewable, finite and limited. Some, like petroleum, are recognized in the national consciousness as being in dwindling supply, and that recognition has cost us dearly in the marketplace.

So we make and light the fires on November afternoons, conscious of the wonder of wood as a renewable resource, conscious of the sun and the wind as contributors to home heating, yet informed by the discovery of barbed wire in an ancient line tree that past keepers of these places were here only yesterday and seldom imagined the consequences of man's arrogance.

Many other wood burners — and I am shamed each winter day by the neat cathedrals of stacked firewood they have built in sheds and under porch roofs — seem able to gather enough wood in September to last them until May.

I have never done that. It is usually March, or before, when the woodpile disappears. Then I go hunting and harvesting driftwood from the frozen beaches. It is a legacy of the winter storms. The salt burns green, corrodes the stove pipe chimney, and starts me pondering on the wonder that I may be burning a bit of Samoan mahogany, a plank from a Genoese barge, or a line tree that marked the burial of Captain Kidd's treasure on Gardiner's Island.

There are more than a score of dreams to every half-cord, no matter what sort of wood you burn, just as there are more than a score of enjoyments that can never be catalogued in the primers about woodstoves and fireplaces.

. (JNC)

21

Heating with Wood— Some Details

I can think of four reasons why I heat my house with wood: pollution, cost, availability, and psychology.

Pollution: Wood is a renewable resource formed using solar energy in the photosynthetic process:

Sunlight + Carbon dioxide + Water = Sugar (Cellulose) + Oxygen

The combustion process is essentially the reverse:

Cellulose + Oxygen = Heat + Water + Carbon dioxide

These two processes go on, independent of man, at the same rate. Whether man combusts wood in a few hours in his stove or whether nature combusts the fallen tree over the years makes little difference. The carbon dioxide added to the atmosphere is the same in each case and will be taken up by the trees in growing next time around. However, the carbon dioxide generated by the combustion of nonrenewable coal and oil reserves represents an additional load on the atmosphere, the results of which are as yet unclear.

A small amount of pollution is produced in the form of oxides of plant nutrient minerals, but with proper combustion the amounts are far less than in the case of oil or coal. And there is no way one would compare the natural pollutants of combustion to the man-made potentially lethal pollution from nuclear power plants.

Cost: Table 20 lists equivalent fuel prices. As an example, oil at 40 cents per gallon makes hardwood worth $80 per cord. Electricity at 2 cents per kwh makes hardwood worth $100 per cord. In Maine, the 1974 price for cut, split, and delivered hardwood was $45 per cord. Pulpwood jobbers will deliver hardwood in 4-foot lengths for $30 per cord in a six-cord minimum load. People who *really* heat with wood wouldn't be caught buying wood any more than a hunter would buy venison. Wood is everywhere for the taking, rotting away at the average rate of half a cord per acre per year.

Table 20. *Equivalent Fuel Prices*

Electric 100% eff. ¢/kwh	LP Gas 80% eff. ¢/gal.	Natural gas 80% eff. ¢/therm	No. 2 Fuel oil 70% eff. ¢/gal.	Hardwood @ 20% moisture 60% eff. $/cord
1.0	18.9	20.5	25.1	50
1.1	20.7	22.5	27.7	55
1.2	22.6	24.6	30.1	60
1.3	24.5	26.6	32.6	65
1.4	26.4	28.7	35.2	70
1.5	28.3	30.7	37.7	75
1.6	30.2	32.8	40.2	80
1.7	33.9	34.8	42.7	85
1.8	35.8	36.9	45.2	90
1.9	37.7	38.9	47.7	95
2.0	39.6	41.0	50.2	100
2.2	41.4	45.0	55.3	110
2.4	45.2	49.2	60.2	120
2.6	49.0	53.2	65.3	130
2.8	52.8	57.4	70.4	140
3.0	56.6	61.4	75.4	150
3.2	60.4	65.6	80.3	160
3.4	67.8	69.6	85.4	170
3.6	71.6	73.8	90.1	180
3.8	75.4	77.8	95.4	190
4.0	79.2	82.0	101	200

Availability: Which brings us to the third reason, availability. Whereas fossil fuels are renewable on a million-year scale, wood renews itself over 20 to 40 years. The Arabs own a lot of oil. I own a lot of trees; and my trees will still be producing when the wells have run dry. Our earlier calculations showed that a well-sited and well-insulated passive solar house of average size could be heated with 2 cords of wood per year. How many houses could be heated with wood?

Larry Gay in *The Complete Book of Heating with Wood* cites the availability of waste wood as shown in Table 21.

The table says that the potential indefinitely sustainable firewood yield east of the Mississippi is 146 million cords per year. This wood is both hardwood and softwood. Therefore, there is enough to heat roughly 50 million passive solar homes in a climate as rugged as Maine's, all without curtailing the paper or lumber industries.

Wood heat will never become the fuel for all homes in the United States because of the transportation cost (a cord of hardwood weighs about 4,000 pounds) and because of concentrated populations. But I hope at least to have convinced you of the availability of firewood if you build in the country.

Psychology: As we have already learned, heat may be transferred by three methods: conduction, convection, and radiation. Nine out of ten heating systems installed today heat by convection alone, either gravity or forced. The ultimate engineering goal seems to be an absolutely uniform air temperature throughout the house. But different people like different temperatures at different times. We haven't all been cast in the same mold. The human body itself generates and dissipates heat at rates dependent upon level of activity. For the same sense of comfort

Table 21. *Potential Pulpwood in the Eastern United States*

Region	Manufacturing Waste	Logging Waste	Fire, Wind, Disease Loss	Net Growth (annual gain)
Northeast	95	177	565	1,630
North Central	77	106	692	3,997
Southeast	167	336	616	1,170
South Central	176	342	554	978
Million cu.ft. (totals)	515	961	2,427	7,775
Million cord	6.4	12.0	30.3	97.2 = 145.9
				total per year

a person requires a lower ambient temperature when active than when passive. The wood stove alone heats by convection and radiation. In my house, natural gravity convection keeps the air temperature uniform within ±5 F° throughout the house. Any spot within line of sight of the stove, however, gets a radiation bonus. The radiant heat is felt to increase as one approaches the stove. Therefore, there is a range of apparent temperature spanning the comfort zone — all without chilling, infiltrating drafts.

I like to think of my stove as a person — the closer I get, the warmer the response. I might point out that radiant heat, introduced to this country by Frank Lloyd Wright, is considered by most architects to be the ultimate form of heating. Its use as Wright envisioned it (floors or ceiling heated from within by electric cables or hot water pipes) has been severely limited by the cost of installation. Few people recognize the wood stove as a radiant heater — if they recognize it as a source of heat at all. Even in Maine, a house heated solely by wood is considered to have no central heating and so is taxed lower!

The Fuel Value of Wood

Beechwood fires are bright and clear
If the logs are kept a year.
Chestnut only good, they say,
If for long 'tis laid away.
But ash new or ash old
Is fit for queen with crown of gold.

Birch and fir logs burn too fast
Blaze up bright and do not last.

It is by the Irish said
Hawthorne bakes the sweetest bread.
Elm wood burns like churchyard mold;
E'en the very flames are cold.
But ash green or ash brown
Is fit for queen with golden crown.

Poplar gives a bitter smoke,
Fills your eyes and makes you choke.
Apple wood will scent your room
With an incense like perfume.
Oaken logs, if dry and old,
Keep away the winter's cold.
But ash wet or ash dry
A king shall warm his slippers by.
— Anonymous

Wood is wood. That is, the major difference between species of wood is not chemical composition but density. The more wood fibers per cubic inch, the stronger and heavier the cubic inch of wood. The fuel value of wood is related only to the chemical composition. Therefore, a pound of wood has about 8,600 BTU potential fuel value independent of species. Table 22 shows the weights and fuel values per cord for common fuelwoods.

The effective fuel value of the wood is reduced by about 1 percent for every 1 percent of moisture content. Air-dried wood or wood stored outside but protected from direct precipitation will equilibrate at about 20 percent

Table 22. *Fuel Value and Weights of Common Fuelwood @ 20% Moisture*

Species	Lb./Cord	Fuel Value/Cord (Million BTU)
White oak	4,400	30.8
Sugar maple	4,100	29.6
Red oak	3,900	27.4
White ash	3,700	26.0
American elm	2,400	23.8
Paper birch	3,400	23.8
White pine	2,200	15.8

moisture content in the winter. Wood stored inside the house before burning will equilibrate at less than 10 percent moisture content in the winter unless the house is humidified. The reduction in effective fuel value is due to the robbing of heat energy to vaporize the water. However, there is a much more powerful negative effect of moisture in holding down the fire temperature that, by preventing complete combustion, gives a lower efficiency and much creosote.

Don't forget that the value in Table 22 must be multiplied by stove efficiency to get BTUs released to the room. Stove efficiency is defined as:

$$Efficiency = \frac{Heat\ added\ to\ inside\ of\ house}{Fuel\ value\ of\ wood} \times 100$$

If the stove efficiency were 60 percent, then 60 percent of the fuel value in BTUs would end up inside the house and 40 percent would leave via the smokepipe.

How Stoves Work

Illustration 116 shows a poorly designed but rather common form of wood-burning stove. It illustrates the process of combustion and the creosote problem common to all stoves. The process of combustion consists of three phases. First, all of the water must be driven from the wood before the temperature of the wood can rise much beyond 212° F. That is why you can't seem to ever get any heat from a wet wood fire. After the water is driven off, other higher temperature volatiles are driven off,

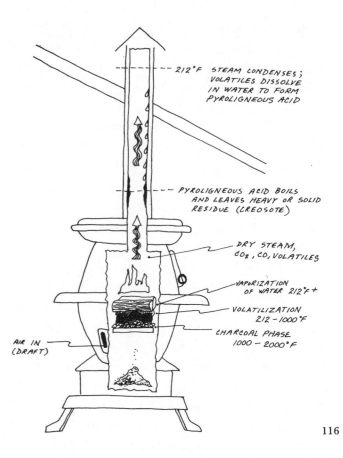

212°F STEAM CONDENSES; VOLATILES DISSOLVE IN WATER TO FORM PYROLIGNEOUS ACID

PYROLIGNEOUS ACID BOILS AND LEAVES HEAVY OR SOLID RESIDUE (CREOSOTE)

DRY STEAM, CO_2, CO, VOLATILES

VAPORIZATION OF WATER 212°F +

VOLATILIZATION 212 – 1000°F

CHARCOAL PHASE 1000 – 2000°F

AIR IN (DRAFT)

116

including methanol. In a fireplace these volatiles can be seen to burn as little gas jets, just like in a gas stove. Visible smoke rising from your chimney consists of unburned volatiles. Complete combustion is represented by mere shimmering heat waves. Therefore, use your chimney output as an indication of combustion. After all of the volatiles have been consumed, we are left with nearly pure carbon, or charcoal. After extinguishing a roaring fire, you will find these "charred" or charcoal sticks. In fact, charcoal is produced commercially by heating wood without oxygen until all of the volatiles are driven off. The charcoal phase represents the creosote-free phase. After oxidation (combustion) of the charcoal we are left with only the mineral content of the wood ashes. Since these are minerals necessary to the growth of trees and are being continuously removed from the soil by all growing things, we would be doing nature and ourselves a favor by returning the ashes to the ground. Years ago, wood was burned in great open pits and the collected ashes were shipped to England to be used as fertilizer!

DRAFT

Just as with convection of heat in buildings, strong convection in a stovepipe depends upon a large ΔT, large vertical distance, and small friction. Convection up a stovepipe is known as the draft. When a fire is burning strongly, the ΔT is very high and little problem is encountered, but in starting a fire we have initially a very small ΔT. A large vertical distance, i.e., a tall chimney or stovepipe, will compensate for this. If a problem is still encountered, holding a burning newspaper up inside the smokepipe will start the draft flowing. Trees near the smokepipe or chimney outlet can cause turbulence in the air during winds. When a turbulent eddy blows down the chimney we have a backdraft, resulting in smoke puffing out of the stovepipe and stove. A "hat," or vaned deflector, will help in this case. If all else fails, the smokepipe must be lengthened or the tree removed. A drafty house or a house subject to high infiltration will sometimes have a lowered internal pressure or partial vacuum. This pressure differential may be impossible to overcome other than by tightening up the house or opening a window or door on the windward or high-pressure side of the house. Of course, you won't encounter this problem when you build your own zero infiltration house!

CREOSOTE

Creosote is known to most wood burners. Sooner or later conditions will be favorable for its formation in most stoves. As shown in Illustration 116, unburned volatiles go up the stovepipe along with live steam. If the temperature in the pipe drops below the condensation point (around 240° F), the steam condenses along with some of the volatiles.

Together they form a brown liquid known as pyroligneous acid, which flows back down the pipe. The water boils off as the acid drops nearer to the fire, leaving an increasingly thick goo. When all of the water is gone, the residue resembles week-old drippings from an apple pie in an oven but with the odor of a burned chicken house. If the stovepipe is installed in the normal way, the creosote runs out of the pipe. Since a stovepipe "draws" its entire length, installing the pipe upside down will solve the external creosote problem and the smoke will (surprise!) never come out any more than it did before.

If you think about Illustration 116 awhile you should be able to predict the factors that favor formation of creosote. Creosote will not be deposited in the smokepipe if: the volatiles are burned within the stove, or unburned volatiles leave the smokepipe before condensing. *Therefore, to prevent creosote:*

● Consider a stove designed to burn the volatiles by a secondary, or even better, downdraft system. Even with such a stove, however, there are bound to be times when unburned volatiles escape.

● Never close the damper (draft control) down on a fresh load of wood. By turning down the draft we are depriving the fire of oxygen needed to burn the volatiles just at the time of maximum volatile release. In addition, the flue gas temperature is lowered, perhaps to the point of condensation within the smokepipe. Instead, let the fire burn robustly until at least the steam phase is passed before turning the draft down for the night.

● Burn dry wood. If you have nothing but wet wood, burn it a little at a time rapidly and finely split.

Burn pine only if completely dry. All wood has creosote potential, but pine, spruce, and cedar have the greatest amounts.

● If the temperature drops to 240° F within the smokepipe, you're in big trouble. To hold the temperature drop down, make the smokepipe as short as possible, make it of low emissivity (bright galvanized instead of black or red, etc.), or as a last resort use Metalbestos or equivalent insulated pipe the entire length. Note that the pipe will lose much more heat running outside a building than inside.

● Use as heavy a gauge galvanized steel stovepipe as possible. It comes ready-made in several gauges. Sheet metal shops will fabricate it from even heavier sheet at minimal cost. The heavier gauge will not only last a lifetime but will be reassuring in case of a chimney fire.

If, in spite of all precaution, you do get a heavy creosote buildup you have a choice: either you clean it out (spiraling a chain from above is a neat method) or chances are nature will do it for you by way of a chimney fire (creosote burn). The old-timers sometimes intentionally let the creosote burn out. If the buildup is not excessive and the *smokepipe or chimney is heavy and sound,* the fire will pose no problem. I do it myself about once a month by opening the stove-loading door while burning a big load of dry pine; with a modern airtight stove the draft can always be shut so tightly that the chimney fire goes out.

Apparatus

FIREPLACES

Chimney fires are not as common in the case of fireplaces as with wood stoves. The fireplace provides plenty of air to burn everything, including the volatiles. The major problem with the fireplace is that it gulps much more air than it needs for complete combustion. Much of the air is simply drawn from the room up the chimney without ever contributing to the combustion process. The other problem of the fireplace is that it heats by radiation only. All of the air that comes into contact with the hottest surface of the fireplace ends up outdoors. Both the Franklin and Heatilator-type fireplaces are improvements that basically bring air into contact with the backside of the hot surface and then release it to the room by convection (Illustration 117).

Due to the extreme amount of infiltration required to operate the common fireplace, its efficiency probably never exceeds 15 percent. If a fireplace is operated in parallel with a central heating system, you may actually be losing ground! That is, the heat supplied by the central heating system in warming the infiltrating air may be greater than the heat gained in

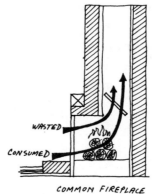

COMMON FIREPLACE   "HEATILATOR"   117

the living room by radiation from the fireplace. This is not a contradiction. In the old days, the folks used to gather up around the fireplace while the rest of the house remained bitter cold due to infiltration. Now we ignore the fact that the furnace is working overtime (unless the thermostat basks in the rays of the fire) and think we're saving money as we throw another log on the fire. If you must have a fireplace, consider the research done by Dr. Franklin and provide the draft directly to the fire from outside the house. Jøtul also makes a "Combi" stove that can be operated open as a Franklin type (with the same low efficiency) or closed with the same high efficiency as the airtight.

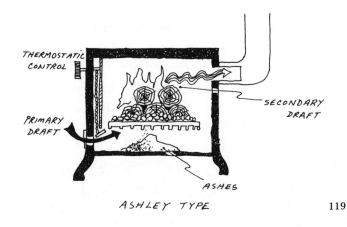

ASHLEY TYPE 119

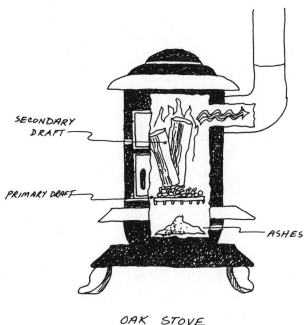

118 OAK STOVE

ANTIQUE STOVES

The value of controlling the draft to a fire was recognized long ago. The famous and beautiful old "oak" stoves (Illustration 118) were the first serious attempt at an airtight design. Heavy sheet metal was used in an attempt to

eliminate unwanted air inlets. Intake was largely confined to the drafts, primary and secondary, so that the combustion process could be controlled. They were largely successful, and an oak stove is prized by many owners for its ability to burn long (overnight if you learn proper care and feeding) and clean. The overall seasonal efficiency, judged from my "round oak," is probably around 40 percent.

ASHLEY TYPE

There are several brands of stove (Ashley, King, Atlanta, and Sears) that are suspiciously similar (Illustration 119). Two of the names have identical mailing addresses. To the eye they are identical. The key feature is a bimetallic thermostatically controlled primary draft that causes a constant rate of heat production in BTU/hour.

To call the draft control a thermostat is a little misleading since it's the stove temperature that is controlled, not the room temperature. The control contributes to a lower and more predictable fuel consumption. That and the ability to hold about 100 pounds of wood make it a reliable 12-hour burner. However, the thermostat is not smart enough to detect creosote production, and improper use of the Ashley has given it a reputation as a creosote producer. Improper use consists of cramming it full of fresh wood and immediately turning the draft down just before bed! As we've seen in Illustration 116, nothing will produce creosote faster than that!

A theoretically more sophisticated design than the Ashley is the Riteway, which forces the volatile gases into close contact with the fire in a true downdraft system (Illustration 120). Although a little more complicated to operate, it should produce less creosote. Riteway also makes a complete line of wood furnaces.

SCANDINAVIAN BOX STOVES

My favorite stove is the Scandinavian airtight box stove (Illustration 121). Four popular brands are currently Jøtul, Trolla, Lange, and Morso. From the outside they look like the old schoolhouse box stove, but looks are deceiving! The Scandinavians have been heating with wood and working on their stove designs while we've considered wood burning funky or quaint. These

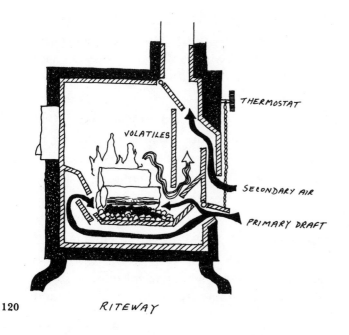

NORWEGIAN BOX STOVE   121

RITEWAY   120

box stoves have been refined to the point where they are as efficient as oil burners.

The primary and secondary drafts are operated together and consist of a series of small holes. As a demonstration of the excess maw of the average fireplace, all of the draft holes together in one of these stoves add up to about 1 square inch! This should also tell you something about how much infiltration is required to heat with wood. The small draft holes are shaped so as to cause turbulence in the incoming airstream. The turbulence causes mixing with the escaping volatile and exhaust gases that have been forced to the front by a heavy metal baffle over the fire. By this means, oxygen is mixed with volatiles in a high-temperature region, promoting combustion of the volatiles.

The baffle plate has two further functions. All of the primary-draft oxygen is consumed immediately in the fire at the front end of the logs. No oxygen is available to support combustion at the rear. In this way the logs are forced to burn like a cigarette. I have tested this notion by throwing a Corn Flakes box to the rear of the stove when the fire was started. Sure enough, 4 hours later the paper was still unburned, even though the fire was roaring. The small draft results in a very intense localized combustion, which progresses extremely slowly to the rear. After the fire is reduced to a few embers at the rear, the whole process is reinitiated by pulling the coals to the front and reloading. The stove is so airtight that closing the draft will extinguish the fire. At the lowest sustainable draft setting the models burning 24-inch logs will burn 12 hours.

A last function of the baffle is to hold the flue gases for a longer time inside the stove, where they give up more of their heat. Efficiency and output curves for the Jøtul stoves are given in Illustration 122. The BTU/hour curves can be used with the heat calculations in Chapter 19 to determine the proper size of stove for your house. The higher the stove efficiency, the lower the temperature of the flue gas. With a stove efficiency of up to 75 percent, the flue gas temperature can be precari-

ously low, so be warned! A lot of people are surprised when they put a long black smokepipe on a Scandinavian stove and get creosote in return. The additional heat you'll get from the pipe isn't worth it!

Heating Geometry

Heating by wood stove is effected by convection and radiation. The strength of *convection* depends upon: ΔT, both vertical and horizontal; height of the open space; and freedom of air current motion. *Radiation* travels by line of sight and so is effective only in areas within view of the stove. Also, the strength of radiation decreases with the distance from the stove. With these qualitative observations in mind, we can choose the best location for a wood stove for various house geometries.

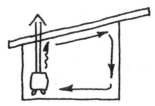

123

OPEN PLAN, SHED ROOF WITH SOUTH WINDOWS

With such strong circulation, I have found in my open 20' × 32' social house that the air temperature at a uniform 4-foot height varies by less than ±5 F° throughout the house.

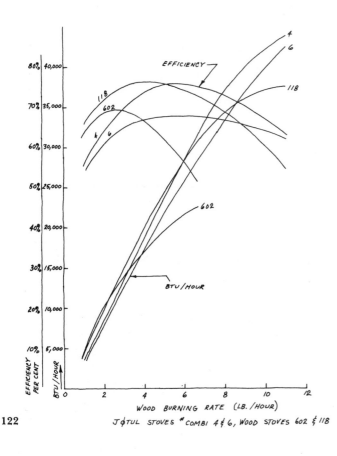

122

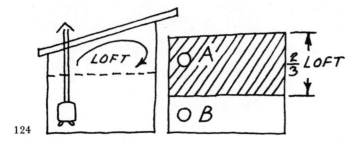

124

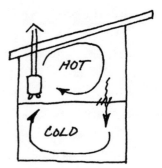

126

STORY-AND-A-HALF (LOFT)

With the stove at A, heat is trapped under the loft. The heat that escapes to the loft is given up in a secondary circulation with the upper windows. Believe it or not, the temperatures upstairs and downstairs are *equal*!

With the stove at B, however, the heat rises directly to the loft. The temperature upstairs now runs 15 to 20 F° higher than downstairs.

However, an exposed shed roof is appealing in a living room. Everyone wants to sit around a living room stove. With the stove up, the upstairs now resembles case A, but the downstairs is cold. Putting registers in front and back as above will result in circulation, but the downstairs will still be at the temperature of the air falling from the windows (cold!).

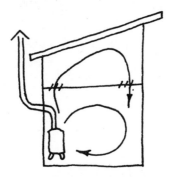

125

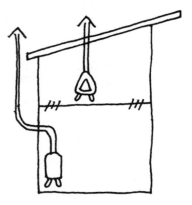

127

TWO-STORY SHED

With the stove downstairs and a solid floor above, the downstairs resembles case A while the upstairs is too cool. By putting registers in the floor we can set up a two-story circulation to any degree we wish. Registers at the front *and* rear walls will encourage a much stronger circulation than simply one at the rear because of the cold air return. For a conventional room arrangement (living room down, bedrooms up), this solution is fine.

The solution when a fire is desired in an upstairs living room is to put the primary airtight stove downstairs and registers in the floor, thus giving complete control of both upstairs and downstairs from the primary heater. In addition, a small, low-efficiency, open-flame stove in the upstairs living room will add little heat but lots of psychological effect. When both fires are operating, the floor registers should be closed.

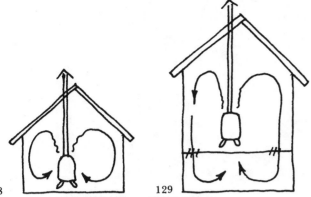

128   129

going to heat themselves. We've really been put to sleep by Honeywell, I guess.

Here are a few ideas.

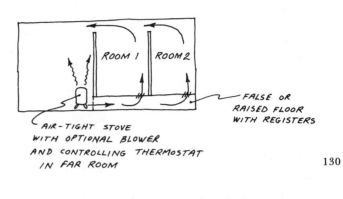

ROOM 1 ROOM 2

FALSE OR RAISED FLOOR WITH REGISTERS

AIR-TIGHT STOVE WITH OPTIONAL BLOWER AND CONTROLLING THERMOSTAT IN FAR ROOM

130

OPEN BUT SYMMETRICAL HOUSE

In the conventional symmetrical house, heat is lost equally by all walls. Therefore, the strongest and most uniform convection will be produced by locating the stove at the center of the house. With a full basement or insulated crawl space, full-width floor registers beneath the windows and under the stove permit a draft-free ("no-draft") floor.

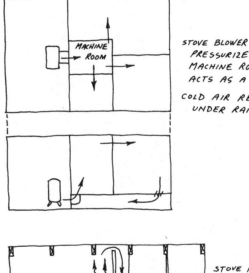

MACHINE ROOM

STOVE BLOWER PRESSURIZES THE MACHINE ROOM WHICH ACTS AS A PLENUM.

COLD AIR RETURN UNDER RAISED FLOOR

131

CLOSED ROOMS

Heat will not flow to closed rooms by either radiation or convection. A common mistake I see in designs for wood-heated houses is the assumption that a lot of little closed rooms are

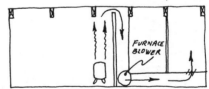

FURNACE BLOWER

STOVE NOT COMPATIBLE WITH BUILT-IN BLOWER

132

22

Wiring, Plumbing, Light, Sound, Ventilation, Etc.

When it comes to electrical wiring and plumbing, time is an essential — unless you also happen to be a professional electrician and plumber. If you are not, you will have to learn how if you want to wire and plumb your shelter. And learning takes time.

We were short of that commodity when we designed and built our place. We had sold our former home with an occupancy deadline — the would-be owner wanted to move in September 1, or else no deal. We took the gamble and gave ourselves just five months to build our new home. (We made it in six, although we had to share the living room with a bench saw for yet another thirty days.)

Under that sort of temporal pressure, neither I nor the relatively unskilled helpers who made up the construction force could take the time to learn the relatively intricate knowledge needed to wire or plumb a house competently and safely. We hired professionals for 80 percent of the work.

One of the young carpenters had done some electrical work during a stint with the U.S. Navy and, after the professionals had installed the main switchbox and circuits, he was able to do the simple wiring involved with our overhead, permanent lights. As an ex-radio operator for the Air Force, I knew enough about electricity to understand what both he and the professionals were doing in principle; I did not, and do not, know how to get the job done in detail. I found it most helpful, however, to be able to comprehend the philosophy of electricity as well as the basics of plumbing. Because I did, I was able to talk more coherently to the professionals, and was not quite so intimidated by them as I had anticipated. Like anyone who speaks a different language, they seemed flattered and pleased that I was trying to learn theirs.

I would advise you to do the same. There is a vast temporal difference between learning the general concepts of wiring and plumbing and becoming skilled enough to implement them with your own two hands. Unless you have a year or more to work on your place, I would not argue on behalf of acquiring those skills.

But no matter how little time you have, you must conquer the concepts. Only then will you have a real enough understanding of the critical anatomy of your home to be at peace with the place.

Then, if anything does go wrong after you move in, you won't be baffled by its mystery or overawed by its complexity. You may still have to phone the plumber (and good luck to you), but at least you'll be able to tell him what's gone wrong, and perhaps why. You'll also be able to recognize the scale of the mishap, and thus won't become victim to the kind of anxious fantasies that might lead you to believe the place will soon vaporize, or be flooded off the map.

A conceptual knowledge also gives you a good leg up when you deal with the plumbing inspector and the electrical inspector, and deal you must, or you won't be able to get by that first huge home-building hurdle: the acquisition of a building permit. Inspectors are fine fellows; they have your safety in mind, but you might find it quite impossible to answer their initial questions if you don't have some notion of the amount of electricity you'll be consuming, and why, or what you are going to do with your wastewater, and why.

"Wastewater," by the way, is a misnomer, a contradiction in terms. Conceptually, water is never wasted; and you might just nail that down as your first lesson in plumbing. Every bit of water that comes into your home goes out again, is purified, recollected, and recycled. There is no "end of the world" beyond the edge of the maps where toilet flushings and sink drainings and shower runoffs can disappear. Such "used" water must be cleaned and restored to its original state. Nature has ways of doing that; your "waste" disposal systems will function best if they concentrate on giving nature a helping hand rather than denying her any

role. Besides, chemical treatment systems don't work too well; incinerators use a great deal of energy; and piping your refuse into your neighbor's field is likely to get you in trouble.

Together, my plumber and I worked out a system for purifying our used water that is compatible with our site, helps fertilize our forest, and in the process, restores the water to its original state of purity. Until we created the system, none like it had ever been used in the state. Now there are scores. That's a good argument in favor of your conceptual education.

There are others. If you and yours are the primary builders of your home, it will help a great deal if you can coordinate with your professional plumbers and electricians, and you will not be able to do that efficiently if you don't comprehend why they do what they do. The person who puts your kitchen together, for example, must be able to relate well enough to the plumber to learn what sort of receptacle to build for the kitchen sink and its piped appurtenances. And you as a designer must know enough about your patterns of living and the location of primary furniture to be able to communicate with your electrician about where outlets and switches might best be placed. It is expensive to discover such needs as an afterthought.

As the boss on the job, you should (unless you are terribly wealthy and generous) be able to coordinate a kind of work-flow chart. If you complete the bathroom floor the day before the plumber arrives, you have not only lost that work to the plumber's need to tear up what you have put down, but you also have lost the time of the person who put down the floor, while s/he waits to do the job over again, this time with poor morale.

A few mismanagements like that and you'll find yourself out looking for a new construction crew. No one who has pounded nails all week can watch the work undone without enduring the sort of agony that leads to flight.

These typical tragedies can be avoided in most cases once the concepts of plumbing and electricity become reasonably clear to you — which is precisely the purpose of the following.

Lighting and ventilation, on the other hand, are not quite such serious matters, nor possessed of such grave consequences. After three years in our home we are still altering and modifying our lighting, and we are enjoying the process. We made "track" lights from No. 10 cans painted black; we used spots here and there; and we buy old lamps we think are handsome at country auctions and then try to make them work to the benefit of the place.

My formula for ventilation is simple. There is no room in our home that is not open to a cross draft. Given the almost perpetual breezes that sweep our location, there are just one or two days a year when I consider the acquisition of a fan. By the time I get organized enough to make the decision, the breeze has returned.

We planned it that way when we selected the site.

(JNC)

23

Wiring, Plumbing, Light, Sound, Ventilation, Etc. —Some Details

Electrical Wiring

Electricity has its own terms, which make it sound more difficult than it really is. An electric current is, in reality, a flow of electrons. Therefore, an analogy to the flow of water is easily made.

(1) Whereas water pressure is measured in pounds per square inch (psi), electrical pressure is measured in *volts*.

(2) Whereas the rate of flow of water is gallons per minute (gpm), the rate of flow of electricity is *amps*.

(3) The friction within a pipe retarding the flow of water is poises; electrical resistance is *ohms*.

The rate of work done or rate of energy consumption by electricity flowing is measured in *watts*.

(Equation 1)   *Watts = Volts* $\times$ *Amps*

           Kilowatt = 1,000 Watts

A kilowatt-hour (kwh) is 1,000 watts used for 1 hour (this is the unit of energy sold by the power company).

CIRCUIT

In order for electricity to flow, we must provide a complete circuit. Illustration 134 shows a simple electrical circuit.

The circuit is complete because there is a continuous path for the electrons (current) to

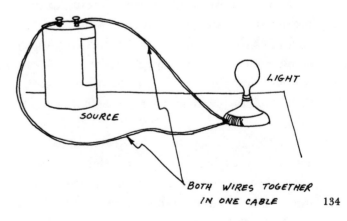

LIGHT

SOURCE

BOTH WIRES TOGETHER
IN ONE CABLE   134

follow, even in the light bulb through the filament. If the circuit is broken at any point, the current stops flowing and no energy is consumed. This is what happens when the light bulb filament burns up and breaks.

The circuit could also be broken intentionally by a *switch* as in Illustration 135. This is how light bulbs are ordinarily controlled. Illustration 135 is really all there is to a "lighting circuit."

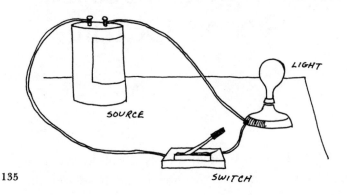

135

RESISTANCE

If we provide no resistance to water flow, there is ordinarily no problem because there is either a fixed amount of water in a tank or there is enough friction in the city water mains to prevent a catastrophic flood. The supply of electrons in the power company lines is so great, however, that we must be very careful to provide resistance in every circuit. Because of the possibility of accidental bypass (short-circuiting) of the circuit resistance, special automatic safety switches are required in all electrical circuits. These safety switches are fuses and circuit breakers.

Every electrical appliance or device consuming electric power has an inherent electrical *resistance,* measured in *ohms.* A very simple relationship exists between volts, amps, and ohms, known as *Ohm's Law:*

(Equation 2)   *Amps = Volts/Ohms*

This relationship is easy to remember if you think of the water analogy. The greater the pressure, the greater the flow; the greater the resistance to flow, the less the flow. Using either Equations 1, 2, or both, it is easy to calculate amps, volts, ohms, and watts, given only two quantities.

AC/DC

In a circuit where the source is a battery, the current flows only in one direction (DC). Reversing the positive and negative signs of the battery reverses the direction of the current flow. For many reasons it is easier, more convenient, and more economical to deal with current that periodically reverses its direction (AC).

Every time the current reverses direction twice, it is said to have completed a cycle. If there are 60 cycles per second, the current is simply known as 60 cycle. In the United States, 60 cycle is standard; in many other parts of the world 50 cycle is standard. In going through one cycle the current actually stops flowing twice. Therefore, there is no electrical current in a light bulb 120 times per second. The reason we don't detect a flicker is the same reason movies seem to move: the human eye smooths out most of the irregularities, not being able to respond to more visual changes than ten per second. Also, the light bulb doesn't completely cool down in such a short time period.

WIRE

Just as a larger-diameter pipe can carry more *gpm* at a given *psi,* a larger wire will carry more *amps* at a given number of *volts.* Another way

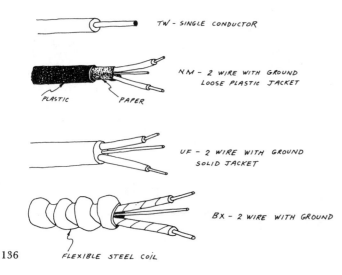

TW – SINGLE CONDUCTOR

NM – 2 WIRE WITH GROUND
LOOSE PLASTIC JACKET

PLASTIC   PAPER

UF – 2 WIRE WITH GROUND
SOLID JACKET

BX – 2 WIRE WITH GROUND

FLEXIBLE STEEL COIL

136

of stating the same thing is to say the pipe and wire both have a smaller resistance. The resistance to flow in a pipe is independent of the material of the pipe because the flow is actually through a hole. However, the flow of current in a wire is through the solid wire. Therefore, the resistance of a wire depends upon both size and material. Copper wire is used for inside house wiring. Because of its lower cost, aluminum is used more and more outside the house. The use of aluminum wire inside the house is not recommended (and in some building codes prohibited) because of its tendency to form junctions of high electrical resistance and therefore overheat.

Cable: Although wire can be bought as single conductors, it most often is used within a jacket as a cable.

Type NM (nonmetallic sheathed cable) is a loose assembly of two or more plastic-insulated wires with or without a bare ground wire loosely sheathed by a light plastic jacket. This is the easiest and most common form of cable used.

Type BX (flexible armored cable) is specified by code in some localities and is the same as NM except the jacket is a flexible metal coil.

Both NM and BX are suitable only for indoor, dry locations. Type UF (underground feeder) is an extremely tough waterproof and sunlight-resistant cable with solid plastic sheathing suitable for use indoors, outdoors, and even underground.

Conduit: Most localities allow residential wiring with one of the above cable systems. However, a few require the use of some type of conduit. There are four basic conduit options: thinwall steel, rigid aluminum, rigid plastic, and surface raceways. All are essentially the same as plumbing systems into which single conductors are fed. They are relatively difficult and expensive compared to the cable systems. Check your local code to find which systems are acceptable.

The resistance of a wire causes a rise in temperature in the wire increasing with *amps.* The safe current-carrying capacity of a wire is known as its *ampacity* and is dependent upon its insulation, proximity to other wires, and whether it is confined or exposed to free air. Table 23 lists ampacities of various common systems.

Table 23. *Ampacity of Copper Wire*

Wire Size	In conduit or cable		In free air		
	TW	THW or RHW	TW	THW or RHW	Weatherproof
14	15	15	20	20	30
12	20	20	25	25	40
10	30	30	40	40	55
8	40	45	55	65	70
6	55	65	80	95	100
3		100 (SE)			
0		150 (SE)			
3/0		200 (SE)			

BOXES AND COVERS

All electrical connections must be made within boxes, either steel or plastic. The boxes must be covered but mounted in locations accessible for making later wiring changes. There are a great number of different sizes and types of boxes available for ease of installation in different ways. A trip to your electrical supply house will show you all the options available. There are six basic shapes available (Illustration 137). The variety comes from the different sizes and method of mounting.

The cable (or conduit) enters the boxes through any of the many "knockouts," which are easily removed with a screwdriver. The cable must be clamped firmly to the box using either the built-in clamp (NM only) or a separate connector.

The cover is chosen from the large number especially designed for the device mounted in the box. If only wire splices are contained in a box, a plain cover plate is used. Illustration 138 shows some of the cover plates available.

138

SWITCHES

There are three different switches that look identical from the outside but are entirely different inside.

Single-Pole Switch: The single-pole switch serves to make or break the connection in a single wire and can control the flow of electricity to any device connected to that wire. Such a switch is identified by having *two* screw terminals.

Three-Way Switch: A three-way switch has *three* screw terminals and is used in pairs to control a light or other device from two different locations (e.g., top and bottom of staircase).

Four-Way Switch: The four-way switch is used to control a light or other device from three locations. It has *four* terminals. *Warning,* there is another switch — the two-pole switch — which has four terminals and is essentially a pair of single-pole switches controlled by one handle.

Dimmer Switch: The dimmer switch is capable of controlling voltage as well as turning

JUNCTION BOXES

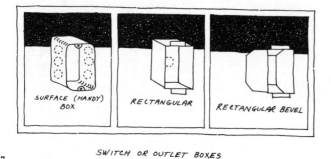

SWITCH OR OUTLET BOXES

137

a circuit completely off. It is used only for controlling incandescent lights. Never use it as a motor-speed control or you'll burn out the motor!

OUTLET RECEPTACLES

Outlet receptacles are the means by which the wires of an appliance are temporarily connected to the wires of the house wiring system. For each wire in the circuit there is a prong and a hole. All new construction must use a system of wiring with a continuous, separate grounding wire. This wire is continuous from the service entrance and may never be interrupted by a switch. Therefore, all cables and outlets must have at least three wires and three holes.

Older houses were sometimes wired without the third wire and therefore use outlets with only two holes. In rewiring older houses never install a three-hole receptacle unless there is a continuous grounding wire that can be traced back to the service entrance lest you mislead someone.

The latest (1975) National Electric Code also requires ground fault interrupt devices (GFI), which automatically disconnect any circuit in which any current is flowing in the safety ground. The reason for such devices will be seen when we discuss grounding. These devices can be plugged into existing outlets to protect only that outlet or installed to protect one entire circuit. They must be used for bathroom and outdoor outlets.

While there are only two types of 120-volt outlets, there are a great number of 240-volt outlets, depending upon the amp rating. Most 240-volt stationary appliances have standardized plugs and receptacles. Illustration 139 shows the two 120-volt receptacles and some of the most common 240-volt receptacles.

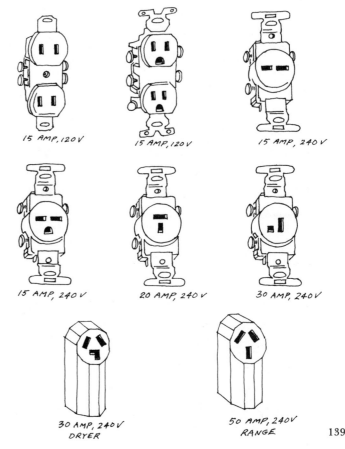

15 AMP, 120V   15 AMP, 120V   15 AMP, 240V

15 AMP, 240V   20 AMP, 240V   30 AMP, 240V

30 AMP, 240V
DRYER

50 AMP, 240V
RANGE   139

120/240 VOLT

Recalling the definition of the rate of power consumption — *Watts = Volts × Amps* — it is obvious that the way to increase the number of watts without increasing the number of amps is to increase the volts. However, increasing the number of volts makes a circuit more dangerous. A very clever trick is employed by the power company, which effectively doubles the voltage without doubling the danger. Notice that there are three wires entering your house. Only one of the wires is at zero, or ground, voltage. Both of the others are at 120 volts but out of phase. Illustration 140 shows the wires and their relative voltages.

Remember that the voltage is AC — that is, the voltage on line A is completely reversing its

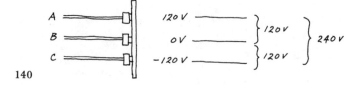

sign or polarity sixty times per second. So is the voltage on line C. But, at any moment, lines A and C are always of opposite polarity. Therefore, the magnitudes of the voltage differences are:

A–B	120 *V*
B–C	120 *V*
A–C	240 *V*

Some appliances run on 240 volts. Others, such as electric ranges and dryers, use both 120 volts and 240 volts. If we run appliances on circuits using three conductors, which are connected at the service entrance to lines A, B, and C, we are supplying the appliance with both 120 and 240 volts.

COLOR CODING

As a safety consideration, the insulation on residential wiring is color-coded. The color code that must be observed is:

White — Neutral or 0 volts (can be 120 volts in some 240-volt systems such as electric heat)

Black — Hot (used in 120-volt circuit)

Red — Second hot (used in 240-volt circuits)

Bare — Second, or safety ground

Green — Connected to chassis or case of appliance or other equipment

THE SERVICE ENTRANCE

The whole service entrance includes the service drop from the poles, the electric meter, and the service entrance or circuit breaker box. Electricity may be desired before the building is ready for the permanent installation. In such a case a temporary service entrance is usually installed on a nearby pole or tree with a few receptacles for extension cords. Illustration 141 shows the specifications for a typical temporary service. These specifications should be obtained from the local utility. Try to borrow or rent a temporary service from someone who has just built a house. You may wish to install the permanent service entrance as soon as possible because the electric rate is usually higher on the temporary meter.

Illustration 142 shows the specifications for a typical permanent service entrance. Again, obtain exact specifications from the local utility. The utility employs a large number of people in customer relations areas. A phone call will instantly produce a multitude of smiling gentlemen in yellow hard hats whose sole purpose in life apparently is making the use of electricity fun and easy. Regardless of their motivation, once you've decided you want electricity, accept

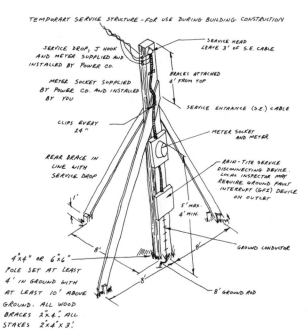

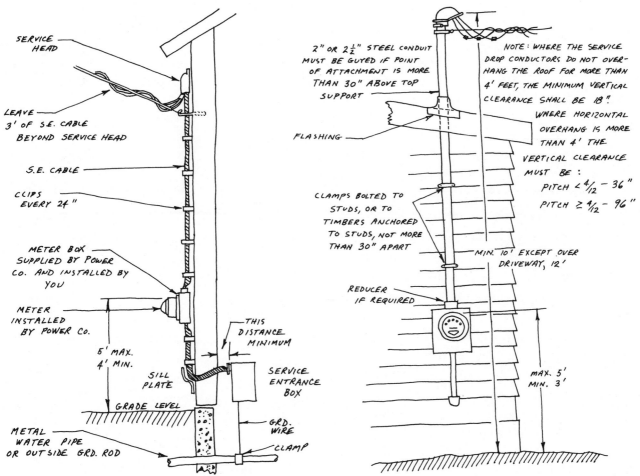

SERVICE HEAD

LEAVE 3' OF S.E. CABLE BEYOND SERVICE HEAD

S.E. CABLE

CLIPS EVERY 24"

METER BOX SUPPLIED BY POWER CO. AND INSTALLED BY YOU

METER INSTALLED BY POWER CO.

5' MAX. 4' MIN.

SILL PLATE

GRADE LEVEL

METAL WATER PIPE OR OUTSIDE GRD. ROD

THIS DISTANCE MINIMUM

SERVICE ENTRANCE BOX

GRD. WIRE

CLAMP

2" OR 2½" STEEL CONDUIT MUST BE GUYED IF POINT OF ATTACHMENT IS MORE THAN 30" ABOVE TOP SUPPORT

FLASHING

CLAMPS BOLTED TO STUDS, OR TO TIMBERS ANCHORED TO STUDS, NOT MORE THAN 30" APART

REDUCER IF REQUIRED

NOTE: WHERE THE SERVICE DROP CONDUCTORS DO NOT OVER-HANG THE ROOF FOR MORE THAN 4' FEET, THE MINIMUM VERTICAL CLEARANCE SHALL BE 18". WHERE HORIZONTAL OVERHANG IS MORE THAN 4' THE VERTICAL CLEARANCE MUST BE:

PITCH < 4/12 — 36"
PITCH ≥ 4/12 — 96"

MIN. 10' EXCEPT OVER DRIVEWAY, 12'

MAX. 5'
MIN. 3'

142

their help. They will advise you exactly where to put the meter and the service entrance. If you play dumb, they may even do a small part of the physical labor.

In spite of their anxiety to please, there is one rule they will not stretch. The National Electric Code states that the portion of cable run from meter to breaker box within the house must be as short as possible — period. This cable carries as much as 200 amps and is unprotected by fuse or circuit breaker until it reaches the main switch in the service entrance. Within the house it poses more danger than outside. My neighbor was forced to move his service entrance box 1 *foot* after all of the house circuits had been connected before he could get power.

Portions of a dozen cables had to be replaced at great cost. Have the utility approve the location of your service entrance box and meter before doing any permanent wiring.

Often we want the entrance box at a different location from the nearest point to the last pole. We have three choices: (1) Decide where the entrance box will be early in the design and have the last electric pole placed accordingly. (2) Run the service drop over the roof to a conduit mast. (3) Run the entrance cable along the outside surface of the house to the point of entry. (The code doesn't restrict external runs.)

GROUNDING

Before discussing actual circuits and how they are connected to the service entrance box, we must understand the functions of fuses or circuit breakers, and ground wires.

All circuits consist of at least one hot wire and one ground wire that together provide a complete loop within which current can flow. Remember that wires have safe current-carrying capacities or ampacities above which resistive heating presents a fire danger. Circuit breakers or fuses are devices inserted in the hot line of each circuit at the service entrance box that will break or disconnect the hot line when the ampacity is exceeded. These protective devices are always placed in the hot line and never in the ground line, the reason for which we will see in Illustration 143.

Figure A shows a circuit supplying a motor that has become short-circuited or overloaded and has blown the fuse. There is no voltage on either the hot or ground line and so you are safe in touching either.

Figure B shows the fuse improperly installed in the ground wire. Although the fuse has blown, the hot wire remains hot and so presents a danger.

Figure C shows a motor in which a short circuit has developed from a point within the motor's resistance to the chassis. Some portion of the device resistance (say one-half) still remains in the circuit and so the fuse may not blow. It may take 20 amps to blow the fuse, but it only requires 0.1 amp to kill a person, so the motor chassis may prove lethal even though the fuse hasn't blown.

Figure D shows how the separate bare or green grounding wire that runs uninterrupted from the service entrance to the chassis of any motor or appliance, to the metal box housing any electrical device; and to the grounding lug of any

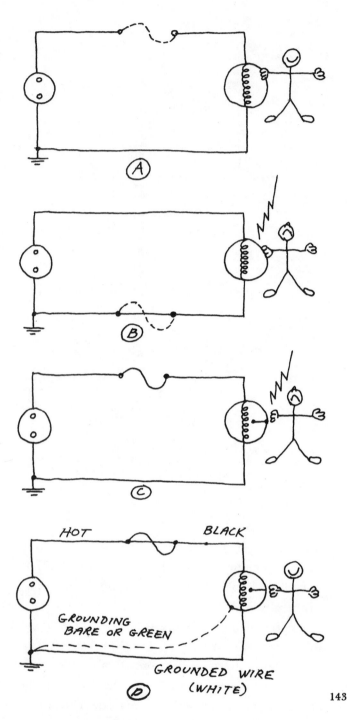

143

receptacle renders the chassis or box safe.

We therefore have the following general rules:

(1) Black and red wires always fused.

(2) White ground wire never fused or broken (as by a switch).

(3) Bare or green wire never fused or broken and always connected to box or chassis.

TYPICAL CIRCUIT AND DEVICES

Circuit devices include: the service entrance box, the main breaker, the circuit breaker, the receptacle, the light fixture, and the switch.

Service Entrance. The box in which the voltage from the service drop is distributed to the multitude of household circuits. It houses both the main breakers and the individual circuit breakers.

Main Breaker. The large circuit breaker or fuse that protects the entire service entrance box. Its amps rating is the same as that of the service entrance.

Circuit Breaker. An electromagnetically and/or thermally activated device that opens an individual household circuit when the rated ampacity is exceeded. Each hot wire leaving the service entrance box is connected to its own circuit breaker.

Receptacle. A conventional device for temporarily connecting a portable appliance or device to a circuit by means of a plug.

Light Fixture. A lighting device permanently wired into a circuit and mounted on a box.

Switch. A mechanical means of controlling a device (usually a light fixture) by means of making and breaking the continuity of a hot wire.

Illustration 144 shows all of the above devices as they operate in a typical 120-volt circuit. Only one of the two 120-volt lines is shown. All of the wires shown are actually in a cable or

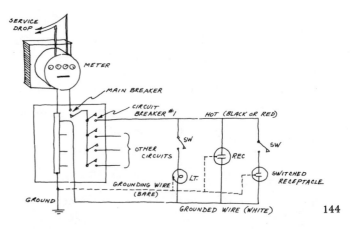

144

conduit together and all connections are made in electrical boxes. Note that the black or red hot wires are the only ones ever switched.

CIRCUITS REQUIRED BY THE NATIONAL ELECTRIC CODE

Lighting Circuits. A circuit capacity of 3 watts/sq.ft. minimum, based upon the outside dimensions of the house. This is potential capacity and not installed lamp wattage. The code also recommends one circuit per 500 sq.ft.

Convenience Outlets, General. No point along a wall, interior or exterior, may be farther than 6 horizontal feet from a receptacle, including sliding glass doors and waist-high (bar) partitions. Sections of wall narrower than 2 feet do not count.

Kitchen and Dining Room. At least two 120-volt/20-amp circuits must be provided only for appliances in the kitchen, dining room, pantry, and family room. Both circuits must appear in the kitchen; only one circuit in the other rooms. The refrigerator and kitchen clock may use these circuits. No lighting may be run from these circuits. One outlet must be provided every

4 feet along every counter space 12 inches or wider. Counters divided by refrigerator, range, or sink are considered separate counters.

Laundry. One 120-volt/20-amp circuit must be provided for the room containing a clothes washer and dryer. The washer may use this circuit; the dryer must have its own circuit.

Ground Fault Interrupt Circuits (GFI). The bathroom must have one convenience outlet near the lavatory (wash basin) protected by a GFI device. In addition, any outdoor receptacle must be protected by a GFI, including temporary services.

Individual Circuits. The rules covering individual heavy appliances are complicated. To be safe, provide a separate circuit for each of the following:

> Kitchen range or countertop plus wall oven
> > 120/240 V, 40 amp up to 8-3/4 kw total
> > 120/240 V, 50 amp over 8-3/4 kw total
> > #8 S.E. cable may be used for 40-amp range and 30-amp dryer
> > #6 S.E. cable may be used for 50-amp range and 50-amp dryer
>
> Dryer
> > 120/240 V, 30 amp regular
> > 120/240 V, 50 amp high speed
>
> Water heater
> > 240 volts/20 amp
>
> Garbage disposal
> > See manufacturer's specifications
>
> Dishwasher (unless portable)
> > See manufacturer's specifications
>
> Furnace
> > See manufacturer's specifications
>
> Water pump
> > See manufacturer's specifications
>
> Permanent electric heat over 1,000 watts
> > See manufacturer's specifications

At this point, I recommend you invest in one or both of the wiring books in the Bibliography. Their pictures are worth a million words.

As a last word of advice, if your town has an electrical inspector, *use him — don't abuse him!* People are usually amazed at the friendly help inspectors provide.

Plumbing

Of all the parts of a house, the plumbing system seems to have the most potential for confounding simplicity. A house is at its most beautiful stage when framed. Immediately thereafter the plumber and electrician usually come in and destroy the simplicity and integrity of the frame by cutting holes, running pipes and wire, and then patching the damage. I won't stand for it in my house. I paid in money and time to create an engineered structure whose frame deserves respect. In the following pages are the essential parts of Appendix B of the Uniform Plumbing Code, which has been adopted by many states and communities; we will come to understand the whys of the code, and investigate a plumbing system that does not violate the integrity of the house shell. Check with your local plumbing inspector to verify his acceptance of the Uniform Plumbing Code.

A house contains two completely separate plumbing systems: *the supply system* (up to the point of water use) and the *waste system* (disposal after use).

THE SUPPLY SYSTEM

The supply system consists of a *water source, pump, building supply pipe, distribution pipes, control valves,* and *fixtures.*

Pumps. If you build in a populated area, your water district probably maintains the water source and pump. In the country you become your own water company. We discussed water sources in Chapter 7 as one of the basic site services. After locating the source we must pressurize the source so that it flows at our command. Pressurizing devices are pumps, often with accompanying pressure tanks, whose function is to avoid having the pump start every time we draw a glass of water. Tanks can also be used as reservoirs for wells with small flows.

The main operational difference between pumps is the depth to which they can be used. The shallow hand pump is practically limited to lifting water about 25 feet since it "pulls" water rather than "pushing." Since the pumping mechanism is entirely within the pump, a hand pump can draw water over unlimited horizontal distances. When drawing long horizontal distances, make sure the foot valve at the bottom of the pipe in the well is operating perfectly. Otherwise the horizontal pipe will empty back into the well overnight and you'll spend half of each morning "priming" the pump. Sears, Roebuck sells a good cheap shallow well hand pump.

Dempster Industries, among others, sells a deep well hand pump. Here the piston is located at the bottom of the well and is connected by a long rod to the handle at the surface. The piston now pushes the water up, so the depth is limited only by your strength. One-hundred-foot depths are not uncommon. This pump may be used on a site with a drilled well but no electricity.

The old reliable piston pump is simply a motorized shallow well hand pump closed at the outlet end so that the piston pulls from well to pump and then pushes the water under pressure into the pressure tank. Since it relies upon pulling water, its water depth is limited to about 28 feet. This type of pump usually requires a small electric motor and doesn't depend upon the motor speed. Therefore, it is easily adapted to small gasoline engines, making it another powerless possibility.

Stepping up in price and down in depth we come to the jet pump. This pump requires two pipes — one for water coming up and one for water returning to the well. It gets a boost from the returning water, which creates an up-pipe pressure by jetting into the incoming pipe in a special jet head. Most knowledgeable people put the maximum depth of economical operation at around 120 to 150 feet. The advantages of this pump over the other deep pump (submersible) are that it is located at the surface where it can be easily serviced (in the house if you wish) and that the only connection to the well is by plastic pipe, making it much less susceptible to lightning damage.

The submersible pump is located, as its name implies, at the bottom of the well. It pushes water and is capable of economically pumping water from depths as great as 500 feet. If your static water level is greater than 100 feet you really have no choice. The problems encountered are: (1) The pump is in the well at the end of a rope, a plastic pipe, and an electrical cable. When the pump fails, you've got yourself quite a project! (2) The pump itself is probably a better electric ground than any ground rod you hammer into the ground. If an electrical lightning surge occurs, the submersible pump is the first place it will head, probably burning out the motor. Several things can improve your average pump life: (1) Submersible pumps with oil motor baths seem to be less susceptible to damage. (2) Lightning arrestors (a simple $20 device hooked across the power feed line) are not foolproof, but are better than nothing. (3) Pull or otherwise disconnect the pump electric feed at the first sign of an approaching thunderstorm and whenever you go away in the summer.

In our little "community" we pump water to three houses from a common pumphouse. Our analysis showed that the cost per house for water from a common pump was 40 percent of the figure for three separate systems. The maintenance cost per family is one-third as much. In addition, if the submersible pump ever fails in January, there are two other sets of hands to share the burden. The best part of all, however, is having the pump and tank out of sight and earshot in their own little beneath-frost house. All of the rust, oil, and condensation fall onto a concrete floor drained to ground instead of the house floor, and valuable heated space in the house is freed for other puposes.

Supply Pipe: The supply pipe runs from the well or other source into the house. I see no reason to ever use anything but black polyethylene pipe. This pipe comes in 250-foot, 100-foot, or cut-to-order lengths in various diameters. Appendix A of the Uniform Plumbing Code contains charts for determining the required diameter from the length of run and water demand. A single-family dwelling with a run over 100 feet should use 1-1/4-inch pipe. For under 100 feet, 1-inch pipe will usually do. Three different pipe strengths are sold, 80 psi, 100 psi, and 140 psi. I always use 100 psi because it doesn't cost much more and I don't enjoy digging up burst pipe. The 140 psi pipe costs a lot more and isn't necessary for horizontal runs. Submersible pumps are sometimes required to pump great vertical distances. The

pressure in a vertical standing column of water increases at about 1/2 psi per foot. Therefore, the portion of pipe in the well should be 140 psi rating up to a depth of about 250 feet. Beyond that depth iron pipe may be required.

Typically, splices and right-angle bends (elbows) will be required in the supply pipe. These joints are made using hollow plastic inserts. Stainless steel hose clamps prevent the inserts from slipping out under pressure. Always use hose clamps in pairs (four per fitting — two at each end) because again, digging up a leaking pipe for the lack of a two-bit clamp is not my idea of relaxation. Why do the joints fail? Imagine a 100-foot-long slug of water traveling at about 10 mph. When your automatic washer closes its automatic valve, the water slams to a stop in about 1/10 second. This is similar to a railroad train hitting a bridge abutment. Except that water is incompressible. Unless your plumbing system has an air cushion, the poor joints get slammed. Hose clamps cost 25 cents.

The inserts are always larger than the pipe. If you're small like me, you have to be clever to get them into the pipe! Dip the poly pipe end in a bucket of hot water and it will stretch easily. Don't use flame because it takes just a few seconds for the pipe to achieve the consistency of roofing tar. This explains why poly pipe is not allowed inside the house by the code.

Distribution Pipe: Once inside the house, water is customarily distributed by piping of either CPVC (chlorinated poly-vinyl-chloride) or copper. You can use whichever you wish. I use copper because I don't like the chemical sound — especially the vinyl-chloride part! The advantage of copper is exactly the disadvantage of CPVC — ease of repair. Most amateur house builders are greeted by several spraying pipe joints the first time they pressurize the system.

With CPVC the faulty joint must be sawed out and replaced by two more. If you're not a good joint maker you could end up with a large number of joints! With copper pipe the joints are simply heated, taken apart, cleaned up, and reassembled for another try.

When a distribution pipe serves more than one fixture (particularly the bathtub or shower), consider using 3/4-inch pipe so that one fixture will not reduce the water flow to another. Some people run separate pipes just for the shower for obvious reasons. Otherwise, 1/2-inch pipe is sufficient.

Freezing: In a crawl-space house, the pipe entering the house is particularly susceptible to freezing. Insulating the pipe doesn't prevent heat loss — it only slows it down. An insulated pipe may go several seasons without freezing because the incoming water at around 40° – 50° F doesn't have time to lose all of the heat required in freezing between replacements. However, if you leave the house for several days in midwinter, chances are it will freeze solid. My pipe has 2 inches of insulation, but it will freeze solid in 8 hours at an outside temperature of 0° F. The solution is a thermostatically controlled "heat tape" available in many lengths at most hardware stores. When the thermostat senses 35° F, electricity is drawn on the resistive wire, producing a low level of heat the entire length. *Caution:* (1) If you insulate the pipe, don't wrap the tape but run it straight alongside the pipe; otherwise, you may accumulate enough heat to melt the tape or the plastic pipe! (2) Use only UL listed heat tapes. (3) Place the thermostat on the pipe — not outside or under the heat tape.

Inside the house I have designed the supply system so that it can be completely drained by closing a pumphouse valve and opening a single valve at the lowest point inside the house, thereby allowing the house to go unheated for any length of time.

On the wastewater side no precautions are necessary ordinarily since the wastewater is at or above room temperature and resides in the pipe for a minute at the most. When I go away in winter, I either drain the fixture traps (the only points with standing water), or pour a cup of alcohol (antifreeze — not Jack Daniels!) down each fixture.

Fixtures: Before designing your bathroom, check plumbing fixture prices. The "architect's Bible" (a Sears or Wards catalogue) will tell you what you'll have to pay to within 10 percent. Unlike other trades, the plumbers have the fixture business all tied up. Plumbing supplies are the only ones the average person cannot obtain "wholesale." If you have a plumber friend, have him purchase your supplies.

When you check the prices you'll find almost immediately that the Chinese Red sunken tub you saw in the *House Beautiful* picture costs about three times as much as the old standby. If you want an unusual bathroom but can't afford those prices, go to a building salvage company, or better yet, offer to relieve a friend of the ugly old lion's-foot cast-iron tub when he modernizes his facility. He'll be jealous and outraged when he finds his old tub sunk into your bathroom floor or standing there with a body of red and toenails of green! You literally could not afford some of the fixtures you can pick up for next to nothing at a junkyard.

THE WASTE SYSTEM

The waste system is required to carry liquids, solids, and even gases — all at the same time and by the natural force of gravity. It is not surprising that the plumbing code is primarily con-

Table 24. *Plumbing Terms*

Branch	any part of piping other than stack
Branch vent	horizontal vent pipe connecting vertical vent to stack
Building drain	lowest pipe to 8 feet beyond building
Building sewer	pipe beyond building drain
Combination waste and vent	horizontal pipe for waste and venting together
Drain	any pipe carrying waste
Fixture	any device discharging waste to drainage system
Septic tank	watertight settling tank for removing solids
Sewage	liquid containing animal or vegetable wastes below
	Human waste – waste water with fecal matter and/or urine
	Gray waste – all liquid waste except human waste
Soilpipe	drainpipe carrying human waste
Stack	vertical pipe for waste and venting
Trap	device providing a liquid waste seal to prevent back airflow
Trap arm	drainpipe between trap and vertical vent
Waste pipe	pipe carrying gray waste only
Wet vent	same as combination waste and vent

cerned with the proper installation of the waste system. A sectional view of a typical waste system, Illustration 145, shows the items defined in Table 24.

Plumbing codes cover the plumbing of all types of structures, including hospitals, hotels, and dormitories. It is easy to get lost among the discussions and rules for installation of ten or more urinals, slaughterhouse drains, etc. Even the plumbing inspector of a nearby city confessed to me that he doesn't understand the whole code. Fortunately, I discovered Appendix B to the Maine State Plumbing Code a separate system for one- and two-story residences. It is short, sweet, and to the point. It is also much less demanding in terms of installation. Check to see if your local code has such a feature.

Table 25. *Fixture Unit Table*

Fixture	Pipe Size (in.)	Units
Toilet	3	6
Lavatory	1-1/2	1
Sink and/or dishwasher	2	2
Shower stall	2	2
Bathtub	1-1/2	2
Washing machine	2	2
Laundry tub	1-1/2	2

SUMMARY OF APPENDIX B TO THE MAINE STATE PLUMBING CODE

Fixture Units: An arbitrary measure of rate of waste flow is the "fixture unit." Table 25 lists the number of fixture units for common household fixtures. The total number of fixture units attached to a drainpipe is a measure of the size of pipe required to handle the expected rate of flow. Note that the drainpipe for a kitchen sink alone, a dishwasher alone, or a dishwasher/ sink combination is the same. Since one is used in lieu of the other, this is sensible. Most sinks come with 1-1/2-inch outlets so a 1-1/2-to-2-inch increaser is required. An automatic clothes washer has a high rate of discharge. When it discharges directly into a standpipe, the pipe must have a diameter of 2 inches. If the washer discharges into a laundry tub, however, the tub requires only 1-1/2 inches. This is because the tub can hold a lot of water and let it out slowly.

Table 26. *Gray Waste Pipe Table*

Fixture Units	Pipe Size (in.)
1	1-1/4
1-1/2 – 8	1-1/2
9–18	2
19–36	2-1/2

Gray waste is all waste other than human waste. Table 26 gives the pipe size (or nominal diameter) required for the number of fixture units draining into it. Pipe sizes are never allowed to decrease from fixture to sewer and

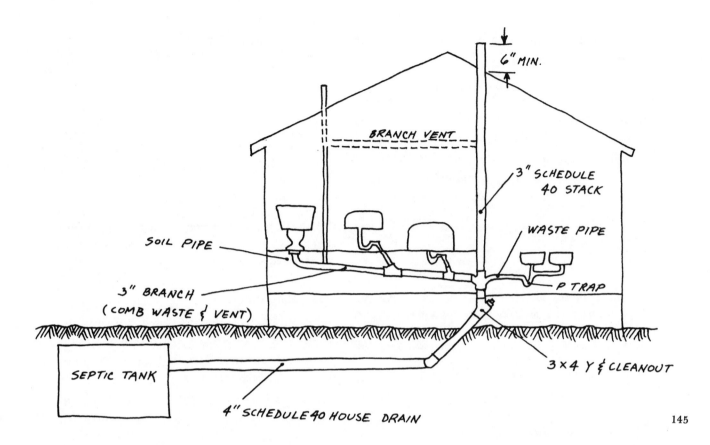

so Table 25 takes precedence. For example, a kitchen sink (two fixture units) requires a 2-inch pipe by Table 25 but only a 1-1/2-inch pipe by Table 26. Since the drainpipe cannot decrease downstream, a 2-inch pipe is called for all the way.

Table 27. *Soilpipe Size*

Fixture Units	Pipe Size (in.)
6–72	3
> 72	4

A soilpipe is a drainpipe carrying human or toilet waste. Since one toilet is assigned six fixture units, Table 27 lists nothing less than six fixture units. If you have more than twelve toilets, use 4-inch pipe! Gray waste and human waste may both travel in a common drainpipe, but, to coin a phrase, "Once a soilpipe, always a soilpipe." That is, any pipe from the point

of entry of human waste is forever after a soil-pipe. It is very common to have smaller gray waste branches empty into the larger soilpipe.

Table 28. *House Drain and Sewer Table*

Fixture Units	Slope 1/8"	Slope 1/4"
6–12	3*	2*
13–24	4	3**
25–50	5	4

* no toilet
** one toilet only

Ultimately, all of the drainpipes empty into the largest and lowest drainpipe in the system — the house drain. Table 28 lists the size of house drain versus total house fixture units and the slope of the house drain. The smaller slope, 1/8 inch per foot, is the minimum required of all drainpipes. The larger slope, 1/4-inch per foot, is the maximum allowed. Even though the pipe is physically identical, the "house drain" changes to the "house sewer" at a point 8 feet from the house.

Changes in Direction: In order to prevent stopped pipes and to facilitate cleaning in case of stoppage, the code specifies methods of changing direction within a pipe. I have found that the rules are easily remembered by making an analogy to a tree. The plumbing fixtures are leaves and the house drain and stacks are the tree trunk. The branches are branch pipes. In a tree:

> The branches always increase in size downward; they sometimes remain the same for a long distance, but they never decrease.
>
> Branches entering from the side are generally smaller in diameter than a main branch.
>
> Branches generally form a "Y" figure, shooting off at about 45 degrees to the main branch or trunk.

In the plumbing system:

> Branch pipes never decrease in size downstream, but may increase.
>
> Horizontal to horizontal and vertical to horizontal connections must be made with 45-degree "Y" branches.

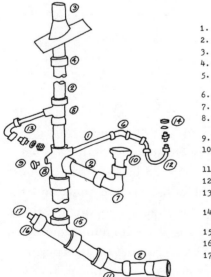

1. 1½" Pipe
2. 3" Pipe
3. Roof Flashing
4. Coupling
5. Sanitary Tee Reducer, 3X3X1½
6. 90° Elbow, 1½"
7. 90° Elbow, 3"
8. Sanitary Tee with two 1½" side inlets
9. Slip Plug
10. Reducing Closet Flange
11. 45° Elbow, 3"
12. P Trap with Union
13. Trap Adapter, 1¼' to 1½"
14. Trap Adapter, 1½" to 1½"
15. Y Branch, 3"
16. Cleanout Adapter, 3"
17. Cleanout Plug, 3"

146

Horizontal to vertical connections must be made with "Y" branches or sanitary tees.

No fitting may be used with more than one branch at the same level unless the discharge from one cannot enter the other. Double sanitary tees may be used only when the barrel is two pipe sizes larger than either branch (e.g., item 8 in Illustration 146).

A cleanout fitting must be provided for every accumulated change of direction of more than 135 degrees.

Changes of pipe direction must be limited to 1/16 (22-1/2 degrees), 1/8 (45 degrees), and 1/6 (60 degrees) bends. *Note:* This rule prohibits 90-degree elbows, but few plumbing inspectors enforce it, provided a cleanout is provided close to the bend.

Traps and Venting: The waste in the septic tank undergoes anaerobic decomposition. Two by-products of this process are methane and hydrogen sulfide. The first gas is combustible, the second smells like rotten eggs. Since the septic tank and drainfield are underground, the path of least resistance is toward the house. To prevent sewer gas from entering the house by way of the drainpipes, every fixture is "trapped." The trap is a device with a low point such that wastewater in the trap prevents the passage of gas. If the flow of water ever completely filled the drainpipe from fixture to stack, we would have created a siphon and all of the water would be pulled from the trap, breaking the water seal. The likelihood of filling the drainpipe full increases with pipe length. Therefore, the trap arm, or length of drainpipe from outlet of trap to inlet of stack, is limited to 10 feet. The vent stack (analog of the tree trunk) serves three purposes: 1) it collects waste from all of the horizontal branches and delivers it to the house drain; 2) it provides air to the branch drains to prevent siphoning; and 3) it provides an outlet for the backflowing sewer gas that otherwise might build up enough pressure to "blow" the trap liquid seal.

Appendix B rules for traps include:

(1) Every fixture must be trapped.

(2) Use "P" traps, not "S" traps.

(3) One trap is sufficient for double and triple sinks and lavatories provided no sink outlet is farther than 30 inches horizontally from the trap inlet or 6 inches vertically from the other sink outlets.

The rules for venting are:

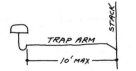

147

(1) The maximum allowed distance from outlet of fixture trap to inlet of stack is 10 feet measured along pipe.

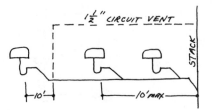

148

(2) *Exception 1:* A circuit or loop vent is allowed provided no fixture trap is more than 10 feet from either the stack or circuit vent. If a separate vent is provided, it must rise as soon as possible in front of the fixture trap.

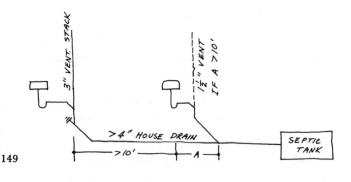

149

(3) *Exception 2:* Not more than three fixtures (not fixture units), not more than 30 inches vertically

nor more than 10 feet horizontally separated, may connect individually to a house drain 4 inches or larger. If the fixture is more than 10 feet from the house drain, the fixture must be vented by 1-1/2-inch pipe.

(4) The vent stack for a house with a toilet must be 3 inches minimum diameter continuous to the termination 6 inches above the roof.

A SIMPLE SYSTEM

At the beginning of the chapter I promised to show how to install plumbing without violating the house structure. Actually, my purposes are greater in number. If at all possible, I would like my plumbing system to:

not penetrate the house vapor barrier except in leaving the building;

be minimal in amount of piping and therefore low in cost;

be installed after all the fixtures are in place;

achieve all of the above without a full foundation.

Illustration 150 shows a house similar to my own in which all of these purposes are achieved. Two tricks are necessary. First, the bathroom floor is raised 16 inches (two steps above the building subfloor). This allows installation of the plumbing after the rest of the building is completed, and the pipe runs under the floor without obstruction. All plumbing is within the heated space of the building. The second trick is Rex Roberts's machine room concept. All whirling or wet machines are located in a room that resembles the boiler room of a ship. No need to camouflage or wall over the plumbing and wiring — the room is exactly what its name implies. The plumbing is forever accessible. The space beneath the bathroom is accessible from the machine room. Backing on the other two

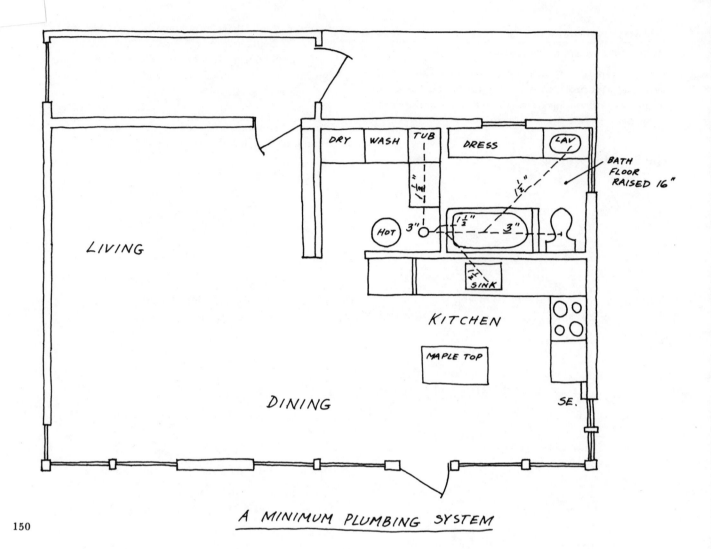

A MINIMUM PLUMBING SYSTEM

150

plumbed rooms — bathroom and kitchen — the machine room concept minimizes piping runs. No drainpipe is longer than 10 feet, measured from the stack, thereby eliminating all circuit vents. The machine room doubles as a pantry, with tens of feet of open storage shelves. I have seen a similar design in which both kitchen and bath floors are raised, achieving a sunken living room effect. A secondary benefit proved to be the huge cool space beneath the kitchen floor for food storage.

THE SEPTIC SYSTEM

The Uniform Plumbing Code requires connection to any public sewer within 500 feet of a house. If you are in a densely populated area, chances are, sooner or later, you'll be required to hook up.

Otherwise you'll require a private sewage disposal system. Of the half-dozen common private sewage disposal systems (trench, bed, chamber, mound, evapotranspiration sand filter, or aerobic treatment system), the type used will be a function of your soil classification and the seasonal high groundwater or bedrock level below the original ground surface.

The largest variable in the true cost of a site is usually the sewage disposal system. Before buying a parcel of land, find out from the plumbing inspector what will be required. If a soil survey is required, get it! At least you will then know the true cost of the land. If the owner refuses to allow the soil survey, he probably already knows the bad news!

Some states are now recognizing composting toilets as a valid way of disposing of human waste and kitchen garbage. Since the toilet accounts for roughly 40 percent of the total liquid waste, the required private disposal system size may be reduced accordingly, but not eliminated. People interested in the composting toilet are often surprised when they find that a regular sewage disposal system is still required for the gray wastes. The pollution potential as measured by the BOD (biological oxygen demand) of gray waste is actually as great as or more than that of the human waste. Therefore, don't expect any major change in the septic system requirement.

Notes on Light

THE EYE AND THE VISUAL FIELD

The receptors in the eye are not distributed uniformly. As a result, the performance of the eye varies with angular distance from the center of vision. The eye can focus on an area only 1° wide, called the visual task area. Beyond the visual task lies the surround, an area in which colors and shapes but not detail are distinguishable. The surround covers ±30°, or a total of 60°. Beyond the surround lies the area of

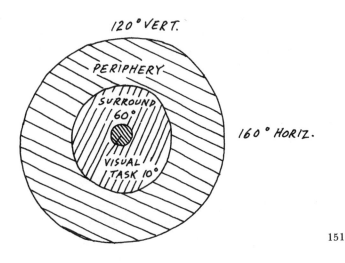

151

peripheral vision, the periphery. In this area, the ability to distinguish colors and shapes is severely degraded.

Illuminating engineers recommend light level contrasts of:

3:1	maximum between task and surround
10:1	maximum between task and periphery
20:1	window to adjacent wall
40:1	anywhere in field of vision

Since the eye has evolved in response to the illumination of the sun and sky, we generally find natural light most pleasing. The incandescent lamp has a spectrum, or light quality, very close to that of the sun. Therefore, in order of visual pleasure we may rank sources of illumination:

Best	daylight
	incandescent
	daylight fluorescent
	other fluorescents
Worst	halogen lamps

Other points to remember are:

Too intense light strains eye muscles (pupil).
Too little intensity lessens visual acuity.

Direct glare is painful.

Indirect or reflected glare is painful.

Natural light consists of about two-thirds direct sunlight and one-third indirect sky radiation. Therefore, the best viewing conditions are with both indirect and direct lighting.

HOW MANY WATTS?

How do we translate the physiological demands of the eye into watts of installed lighting fixtures? The recommended task illumination for various activities is shown in Table 29.

Table 29. *Recommended Illumination (footcandles)*

General background	10
Dining room	20
Kitchen	30
Deskwork	30
Reading	50
Sewing	100

Disregard the definitions of footcandles and lumens — just use the number.

To turn illumination into electrical watts we need a conversion factor between light units and electrical units. This is a measure of efficiency of conversion of energy (see Table 30).

Table 30. *Lamp Lumens per Watt*

Average fluorescent	60
Average incandescent	15
(varies with wattage)	

To calculate direct lighting, assuming a transmission loss of 50 percent, use the formula:

$$Watts = \frac{2 \times Footcandles \times Area}{Lumens\ per\ watt}$$

Example:
$12' \times 16'$ kitchen, fluorescent lighting:

$$Watts = \frac{2 \times 30 \times 12' \times 16'}{60}$$

$$= 192\ watts$$

$$or\ five\ 40\text{-}watt\ bulbs$$

To calculate indirect lighting, we can use the same formula but must account for the effective reflectivity, or percentage of light reflected from the valence and adjacent wall or ceiling.

$$Watts = \frac{2 \times Footcandles \times Area}{Lumens\ per\ watt \times Reflectivity}$$

Example:
indirect lighting for background in a room $15' \times 20'$; fluorescent lamps; average reflectivity of adjacent surfaces judged to be 50% off-white.

$$Watts = \frac{2 \times 10 \times 20' \times 15'}{60 \times 0.5}$$

$$= 200\ watts$$

$$or\ five\ 40\text{-}watt\ bulbs$$

FIXTURES

Good lighting fixtures are expensive. Here are a few money-saving ideas.

In Illustration 152 the last rafter carries half

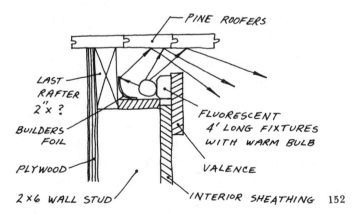

152

the typical load and so leaves a natural alcove for indirect lighting. The valence board trims the wall and shields the bulb at the same time.

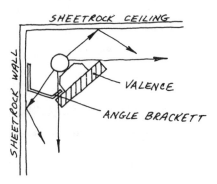

SHEETROCK CEILING

SHEETROCK WALL

VALENCE

ANGLE BRACKETT

153

The arrangement in Illustration 153 is a natural for Sheetrocked walls and ceilings. It gives a remarkable distribution of light. If the wall has texture, such as a rough-sawed board, the texture will be emphasized by the grazing angle of illumination. Try a whitewashed-board wall!

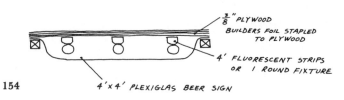

⅜" PLYWOOD
BUILDERS FOIL STAPLED
TO PLYWOOD

4' FLUORESCENT STRIPS
OR 1 ROUND FIXTURE

4'x4' PLEXIGLAS BEER SIGN

154

Illustration 154 shows a fun light panel to use in children's rooms, a game room, bar, kitchen, or over the bathtub. Used or surplus beer signs can be obtained from sign companies for next to nothing. You can leave the paint on or remove it from the clear Plexiglas with lacquer remover. Do not use this light for reading, as the transmission is low.

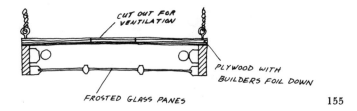

CUT OUT FOR VENTILATION

PLYWOOD WITH BUILDERS FOIL DOWN

FROSTED GLASS PANES

155

Illustration 155 gives the most incredible light I've ever seen. It is an old multipaned window (say 4" × 6" panes) with frosted or stained glass panes. The builder's foil and frosted glass results in a lovely diffused light. Use fluorescents for over kitchen areas or incandescents on a dimmer switch for dining or living room.

If you have a high ceiling (12-foot minimum), you can manufacture your own "track lighting" using 5-inch galvanized stovepipe (Illustration

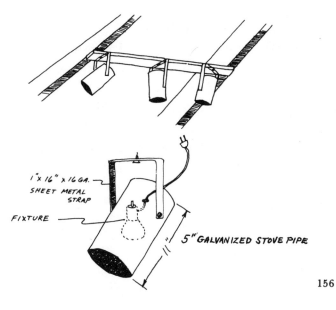

1" × 16" × 16 GA.
SHEET METAL STRAP

FIXTURE

5" GALVANIZED STOVE PIPE

156

156). The gimble mounting (1" × 16" strip) allows the light to be swiveled in any direction, and light is cast upward as indirect lighting and downward as a concentrated direct source. The cost per fixture is about $3. The same lamp in a 4-inch pipe version works well mounted on a wall.

Notes on Sound

MEASUREMENT

The intensity of sound is measured in decibels. The zero-decibel level is defined roughly as the threshold of hearing. A 10-decibel increase is an increase of 10 times in air pressure as detected by an instrument. Since the human ear is non-linear, a 10-decibel increase is sensed as a doubling of volume.

SOUND ISOLATION

The overall (over a wide range of frequencies) effectiveness of a building section (wall or floor, etc.) in preventing the transmission of sound is defined as the sound transmission co-efficient, STC. It is roughly the reduction in decibels achieved by the addition of the barrier.

When two rooms share a common wall, only the wall construction prevents the overhearing of conversation or the aggravation of a whirling and thumping washing machine. We can improve the acoustical performance of a given wall by various inexpensive techniques. Below, you will learn how to build a wall or a floor with any given STC rating.

Table 31. *Typical Sound Levels (decibels)*

Jet aircraft, guns	140
Threshold of pain	130
Inside prop plane	100
Symphony	90
Face-to-face conversation	60
Quiet office	40
Bedroom	30
Whisper	10–20
Threshold of hearing	0

First you must decide what STC rating you are trying to achieve. For any two adjacent spaces specify the probable level of sound at the origin, using Table 31 as a guide. Then specify the sound level desired in the adjacent space. The difference in decibels is the required STC rating of the partition, whether it be a wall or floor.

Table 32. *STC of Basic Constructions (decibels)*

Walls	3-1/2" Stud	6" Stud
1/4" plywood both sides	24	28
1/2" plywood both sides	29	33
1" wood both sides	34	38
1/2" Sheetrock both sides	32	36

Floors	2 × 10 Joist
2" × 10" joist, 1/2" ply-subfloor 25/32" hard-wood finish	25
Same, but add 1/2" Sheetrock ceiling below	37

Table 33. *STC Modifications*

Walls	Decibels
Doubling weight of one side	+3
Doubling weight of both sides	+5
Slotting studs	+8
Staggering studs	+9
Fiberglass fill between studs	+5

Floors	
Resilient clip mounting of ceiling	+10
Subfloor on 1" × 3" sleepers on fiberboard (not nailed) over joists	+10
Staggered floor and ceiling joists	+ 8
Fiberboard (between subfloor and finish floor)	+ 7
Fiberglass fill (only for staggered joists)	+ 3
Linoleum, carpet, cork tile, etc.	0

To compute the STC rating:

(1) Find in Table 32 the rating of the basic wall or floor being considered.

(2) Find in Table 33 the STC improvement due to various modifications to the basic construction.

(3) Add the basic STC, the largest of the modifications, and half of all of the other modifications.

Table 34. *STC of Doors and Windows*

Doors	Decibels
1/2-inch plywood sheet	25
Hollow core door	21
1-3/4-inch solid core door	29
2-1/4-inch solid core door	32
1-3/4-inch hollow core with fiber fill	32

Windows	
Double-strength glass	21
Plate glass	32
Double-glazed double-strength glass	26
Double-glazed plate glass	42

Note: All door STC values assume weather stripping. Without weatherstrip, deduct 10 decibels. In other words, weatherstrip all doors, including the bathroom and machine room.

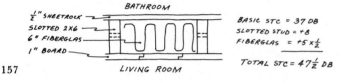

157

Example:

Wall (Illustration 157)
The basic STC rating was found as the average between 6-inch stud walls having both sides Sheetrock and both sides 1-inch board. A slotted stud is one that has been ripped down the middle from 4 inches from the soleplate to 4 inches from the top plate, leaving two practically independent $2'' \times 3''$ studs.

THE QUALITY OF SOUND

The quality of sound in a room is primarily due to uniform response to all frequencies — that is, avoidance of harmonic resonances, and an ideal reverberation time.

Harmonics: One frequency is a harmonic of another if the ratios of the two frequencies form an integer. All musical instruments give off harmonics with every note. The only exceptions to my knowledge are the tuning fork and the synthesizer. If the dimensions of a room such that an integral number of wavelengths just fit between two opposite surfaces, we get a resonance, or apparent amplification. The conventional house is built in such a way as to encourage this aggravating phenomenon by: 1) having all surfaces hard, flat, and parallel to each other; and 2) sometimes having the ratios of length, width, and height such that we get resonance from more than one pair of surfaces at a time (i.e., a room $10' \times 20'$).

Rex Roberts gives the following two rules for avoiding the problem of selective amplification.

• Build the house out of square. This is automatically achieved between floor and ceiling in a shed building. Since these are the largest surfaces, this simple roof will cure most of the problem. Building the outside walls out of square is expensive and time-consuming. However, any interior walls are easily angled a few degrees with minimal extra effort. Angling

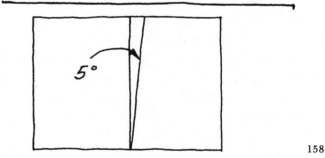

158

one wall cures two rooms! Music practice rooms in music schools are nearly always built this way.

• Compute building dimensions as shown in Illustration 159.

Reverberation Time: With too long a reverberation time, past sounds interfere with present sounds, confusing the ear and tiring the brain.

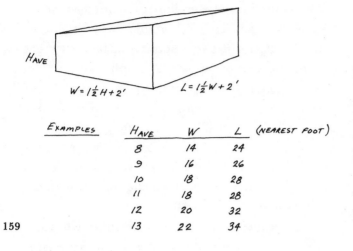

H_{AVE}

$W = 1\frac{1}{2}H + 2'$ $L = 1\frac{1}{2}W + 2'$

EXAMPLES	H_{AVE}	W	L (NEAREST FOOT)
	8	14	24
	9	16	26
	10	18	28
	11	18	28
	12	20	32
	13	22	34

159

Shrill and metallic sounds become very irritating.

On the other hand, too short a reverberation time makes a space "dead." Such spaces are psychologically depressing. The only reason hi-fi demonstration rooms get away with heavy sound-deadening materials such as acoustical-tile ceilings and heavily curtained walls is that most records have artificial reverberation electronically added to make any space sound like a concert hall.

The solution for a house is not acoustical tile and heavy drapes. An ordinary number of rugs, furnishings, and wall patterns such as bookshelves, coupled with a lot of rough-sawed board walls and a shed roof, usually results in a very pleasing reverberation time.

Notes on Ventilation

Ventilation may be achieved by two completely different nonmechanical means: wind pressure and stack effect.

A house located on a site with a strong prevailing summer wind should utilize the natural wind. The wind speeds achievable in the house may be high enough to effect a direct lowering of the comfort index (Chapter 5). Such sites are bare hilltops, bare windward slopes, shorefront (lake or ocean), and wide-open spaces. The ventilating effect of the wind may be increased by:

(1) orientation of openable windows toward the prevailing wind (casement or side-hinged windows allow angling of the building to the wind since the windows can scoop the breeze just as in an automobile);

(2) using shrubs and trees to create high windward and low leeward pressures;

(3) for a given opening area, making the outlet twice the size of the inlets;

(4) locating inlets and outlets at or below the living zone, thus keeping the breeze on the occupants instead of over their heads;

(5) minimizing the interior partitions, permitting a smooth, low-friction flow of air.

Illustration 160 demonstrates how ventilation is modified by shrubs and trees.

It has been found for a house with no surrounding shrubs or trees, and ventilating windows facing directly into the wind, that the wind speed through the window is a fraction of the wind speed outside the house. Equal openings windward and leeward result in the following relation:

$$B = 0.6W$$

where $B = $ *breeze at the window (mph)*

$W = $ *outside wind speed (mph)*

To find the speed of the ventilating breeze inside the house, multiply B by the ratio of

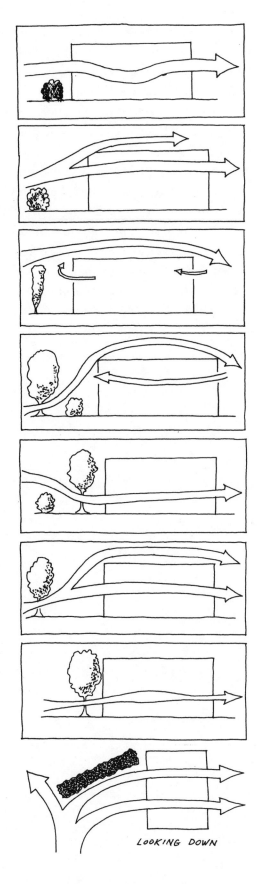

LOOKING DOWN

window opening to effective room cross section. *Example:* outside breeze 5 mph, window opening 20 sq.ft., effective room cross section 8′ × 10′ = 80 sq.ft. Then

$$B \; inside \; room = \frac{20 \; sq.ft.}{80 \; sq.ft.} \times 0.6 \times 5 \; mph = 0.75 \; mph$$

VENTILATION BY THE STACK EFFECT

When there is no strong prevailing breeze to capture, the buoyancy of the inside hot air versus the outside cooler air can be used to move air vertically by the "stack effect." The slightest breeze, however, can overcome the stack effect if the openings are high and low on the walls, so the placement of ventilators relative to prevailing breezes is extremely important. Otherwise, the two effects may work against each other.

For the stack effect the formula is:

$$B = \frac{\sqrt{H \Delta T}}{10}$$

where B = *the breeze at the smaller opening (mph)*

       H = *the height difference (in feet) between the inlet and outlet*

     ΔT = *the temperature inside at the high outlet minus the outside air temperature*

Example:

If $H = 10 \; feet$

              $\Delta T = 10° \; F$

Then, $B = 1 \; mph$

As before, for an effective room cross section of 8′ × 10′ = 80 sq.ft. and openings of 20 sq.ft., B' (felt breeze) = 20/80 × 1 mph = 0.25 mph.

It is important to note that in a well-insulated house, this magnitude of ventilation will quickly decrease ΔT, which will, in turn, decrease B.

160

WIRING, PLUMBING, LIGHT, SOUND, VENTILATION, ETC. — SOME DETAILS / 229

Therefore, ventilation by the stack effect primarily cools by air exchange rather than by creation of a felt breeze.

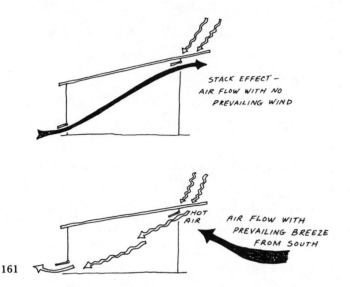

STACK EFFECT — AIR FLOW WITH NO PREVAILING WIND

HOT AIR

AIR FLOW WITH PREVAILING BREEZE FROM SOUTH

161

A specific danger observed in shed buildings with overhangs is shown in Illustration 161.

The roof overhang generates and traps hot air, particularly if the roofing material is dark. A small breeze from the south sometimes reverses the stack effect, driving the hot air into the house!

Whenever available, direct wind-pressure ventilation is most satisfactory. I recommend one front and one rear floor- or sitting-level ventilator *in each living area, placed so that the air movement is directly across the occupants.* First floors and lofts require their own separate vents. Both front and rear ventilators are required to achieve cross ventilation. Don't forget to use a door if it happens to be in the right location.

(CW)

Afterword

It seems like a great deal to do, doesn't it, especially when everything you must do is listed in such detail. The book is like a keg of nails. You might be discouraged if you were presented with a barrel of nails and told to hammer every one into a large pine block.

When you build a house, you'll pound several kegs, and when you finish, you'll hardly remember a single one. Spread over the days, nails aren't that memorable; measured against the glory of your own shelter, they evaporate like dewdrops in the midday sun.

You may live for a year with the travail of shelter building; you will live for a lifetime with the blazing knowledge of your accomplishment. Your home will warm your soul and sustain your self-confidence for the rest of your days; it is a canopy of fulfillment as well as a roof over your head.

And you can unfurl that canopy. We did; and if we did, anyone can.

(John Cole and Charlie Wing)

Bibliography

CHAPTER

1. Turner, J. F. C., and R. Fichter, eds. *Freedom to Build.* New York, Macmillan, 1972.

3. Weidhaas, Ernest R. *Architectural Drafting and Design,* 2nd ed. Boston, Allyn & Bacon, 1972.

5. Banham, Reyner. *The Architecture of the Well-Tempered Environment.* Chicago, University of Chicago Press, 1969.
   Climatic Atlas of the United States. Washington, Superintendent of Documents, U.S. Government Printing Office, 1968.
   Olgyay, Victor. *Design with Climate.* Princeton, Princeton University Press, 1963.
   United States Department of the Interior, National Park Service. *Plants, People, and Environmental Quality.* Washington, 1972.
   Waugh, Albert E. *Sundials, Their Theory and Construction.* New York, Dover, 1973.

7. American Association for Vocational Instructional Materials. *Planning for an Individual Water System.* Athens, Georgia, Engineering Center, 1973.
   Clews, Henry. *Electric Power from the Wind.* East Holden, Maine, Solar Wind Company, 1973.
   Gibson, U. P., and R. D. Singer. *Water Well Manual.* Berkeley, California, Premier Press, 1971.
   Permanent Logging Roads for Better Woodlot Management. Upper Darby, Pennsylvania, United States Department of Agriculture, Forest Service, 1973.

9. Parker, Harry. *Simplified Engineering for Architects and Builders,* 5th ed. New York, Wiley Interscience, 1975.

11. *Everything You Wanted to Know About Plywood.* American Plywood Association, Tacoma, Washington.
    Wood Handbook. Agriculture Handbook Number 72, Forest Products Laboratory, Forest Service, United States Department of Agriculture, Washington, D.C., 1974.

15. Anderson, L. O. *Low-Cost Wood Homes for Rural America.* Agriculture Handbook Number 364, Forest Service, United States Department of Agriculture, Washington, D.C.
    Roberts, Rex. *Your Engineered House.* New York, Evans, 1964.

17. Merrilees, Doug, and Evelyn Loveday. *Low-Cost Pole Building Construction.* Charlotte, Vermont, Garden Way, 1975.

19. Keyes, John. *Harnessing the Sun,* Denver, Conestoga Graphics, 1974.

21. Gay, Larry. *The Complete Book of Heating with Wood.* Charlotte, Vermont, Garden Way, 1974.

23. Richter, H. P. *Wiring Simplified,* 31st ed. Minneapolis, Park Publishing, 1974.
    Simplified Electrical Wiring. Sears, Roebuck and Company.

General Carpentry

Wagner, Willis. *Modern Carpentry.* South Holland, Illinois, Goodheart-Willcox, 1973.

Codes

Basic Building Code. Building Officials Conference of America, 1313 East 60th Street, Chicago, Illinois 60637.

National Building Code. American Insurance Association, 85 John Street, New York, N.Y. 10038.

National Electrical Code, 1975. National Fire Protection Association, 60 Batterymarch Street, Boston, Massachusetts 02110.

National Plumbing Code Illustrated (Uniform Plumbing Code). Manas Publications, Short Tower No. 205, 1868 Shore Drive South, St. Petersburg, Florida 33707.

Southern Standard Building Code. Southern Building Code Conference, 3617 Eighth Avenue South, Birmingham, Alabama 35222.

Uniform Building Code. International Conference of Building Officials, 5360 South Workman Mill Road, Whittier, California 90601.

Metric Conversion Table

Units of	Multiply	By	To Get
length, area, volume	inch (in.)	2.54	centimeters (cm.)
	sq.in.	6.45	sq.cm.
	cu.in.	16.4	cu.cm.
	foot (ft.)	30.48	centimeters (cm.)
	sq.ft.	0.0929	sq.m.
	cu.ft.	0.0283	cu.m.
	yard (yd.)	0.9144	meters (m.)
	sq.yd.	0.836	sq.m.
	cu.yd.	0.765	cu.m.
mass	pound	0.4536	kilogram
	ounce (liquid)	29.57	cu.cm. (liquid)
time	all units identical		
flow	cu.ft./min.	472	cu.cm./sec.
	gallon/min.	0.063	liters/sec.
speed	ft./sec.	30.48	cm./sec.
	ft./min.	0.508	cm./sec.
	mi./hr.(mph)	1.61	km./hr.
pressure	lb./sq.in.	0.068	atmospheres
		5.17	cm. mercury @ $0°C$
energy	British Thermal Unit (BTU)	252	calories
		1,054	watt-seconds
		0.000293	kilowatt-hours
power	BTU/hour	0.000293	kilowatts
	horsepower	746	watts
temperature	F degree	5/9	C degree
	degrees F = 32 + 1.8 (degrees C)		

Index

Vegetation, 30, 41. *See also* Plants; Trees
Vent stack, 220, 221, 222
Ventilation, 117; controlled, 170; cross draft, 203; opening in roof, 151; by stack effect, 229–230; by wind pressure, 228–229
Ventilators, 230
Venting, plumbing, 220–221
Vertical distance of movement, 166
Vertical lifting force, 127
Vertical wind load, 70
Visual task area, 223
Volatiles, 192, 193, 194, 196
Voltage, of appliances, 209–210
Volts: defined, 205; increasing, 209

Walls: in A-frame, 141–142; applying siding, 100; building, 136–137; dead load, 68; fastening rafters to, 150; functions of, 144; how to build, 147–149; internal, 137–138; and joists, 82; side, and overhang, 151; siding and backsplash, 126; and solar heat, 35; STC, 226; telephone poles in, 143; and ventilation, 228; and wind resistance, 71–72; window walls, 72–73, 137–138, 155, 173–175
Warping, in wood, 99, 102
Washer, automatic, 120
Waste: and house building, 5, 76; human and gray, 219; plywood, 103. *See also* Sewage disposal
Waste system, plumbing, 217–221
Wastewater, 202, 217, 220
Water: avoiding trap, 111; and concrete, 122; drainage, 53–54, 120, 125, 126–127, 128, 129; in firewood, 191–192; flow, compared to electricity, 205; and forest microclimate, 44; frost (*see* Frost); and land selection, 51; plumbing

system, 214–223; purifying used, 202; quantities for home use, 56–57; and roads, 53–54; types of, and wells, 55–57; well-water temperature, 123, 128; and wood, 75
Water pipes, 124, 216
Water table, 50, 55, 116, 122, 127
Water vapor, 158. *See also* Vapor barrier
Watts: defined, 205; increasing, 209; lighting needs, 224
Waugh, Albert E., *Sundials, Their Theory and Construction*, 46
Weather, 29, 149. *See also* Climate; *specific entries*
Weathering, of doors, 152
Weather-stripping, 152, 158, 227
Wells, 55–57; pumps, 215
Western frame, 141
Wind, 27; and condensation, 158; cross-draft ventilation, 203; as energy source, 57–58; in forest, 42–43, 44; and foundation structure, 120, 122, 128; and heating, 162, 163, 168; load, 68, 70–72; macroclimatic, 39–41; pressure and overhang, 150; as renewable resource, 186, 187; ventilation by pressure, 228–229, 230
Wind rose, 39–41
Windmills, 57, 162, 163, 186
Window walls, 137–138; cost of, 155; in passive solar house, 173–175; racking in, 72–73
Windows: alignment, 19; in domes, 143; functions, 155; glass for, 159; glazing, 156, 172; heat loss, 152, 172; installation, 157–159; insulating, 156, 157–159; multipaned, 225; nonopening, 155, 170; openable, 228; placing, 137, 142, 155–156, 163; south-facing, 28, 137, 162,

163, 172, 174, 182, 197; south, tables, 179–182; STC, 227; window ratio, 174
Windowsill, 155–156, 157
Winter: and design of house, 11; design temperatures, 30; and forest, 43–44; and heating by windmill, 162; solar heating in, 39, 173–181 *passim*; wind reduction, 43, 162, 163
Wire, electrical, 206–207
Wiring, 201; accessibility, 17–18; color coded, 210; in new housing, 209; with slab foundation, 129; types of wire, 206–207
Wood, 10; analysis, 86, 97; approximating arch, 141–142; bending, different species, 81; as conductor, 166; defects in, 100–102; design cases, 83–86; forces in, 79–86; framing devices, 111; as fuel, 29, 120, 171–172, 182, 183, 185–187, 189–192, 194; "green," 86, 109; nailing, 107, 108, 109; and plywood, 102–103; pulpwood, table, 190; rot in ground, 120; sawing, 98; seasoning, 99–100; shrinkage, 100, 141; span tables, 86, 87–97; stoves and fireplaces, 162, 192–197; values and characteristics of, 75–77; for window frame, 157, 159
Wood stoves, 162, 163, 191; annual fuel requirement, 171, 172; locating, 197–199; operating, 192–194; types, 192, 195–197
Work-flow chart, 202
Workroom, placing, 137
Wright, Frank Lloyd: *The Natural House*, 130; and radiant heat, 191

Yurts, 142, 143